AN ARTIFICIAL INTELLIGENCE APPROACH TO TEST GENERATION

THE KLUWER INTERNATIONAL SERIES IN ENGINEERING AND COMPUTER SCIENCE

KNOWLEDGE REPRESENTATION, LEARNING AND EXPERT SYSTEMS

Consulting Editor

Tom M. Mitchell

Other books in the series:

An Artificial Intelligence Approach to VLSI Design. T. Kowalski. ISBN 0-89838-169-X.

An Artificial Intelligence Approach to VLSI Routing. R. Joobbani. ISBN 0-89838-205-X.

AN ARTIFICIAL INTELLIGENCE APPROACH TO TEST GENERATION

by
Narinder Singh
Stanford University

KLUWER ACADEMIC PUBLISHERS
Boston/Dordrecht/Lancaster

Distributors for North America:
Kluwer Academic Publishers
101 Philip Drive
Assinippi Park
Norwell, Massachusetts 02061, USA

Distributors for the UK and Ireland:
Kluwer Academic Publishers
MTP Press Limited
Falcon House, Queen Square
Lancaster LA1 1RN, UNITED KINGDOM

Distributors for all other countries:
Kluwer Academic Publishers Group
Distribution Centre
Post Office Box 322
3300 AH Dordrecht, THE NETHERLANDS

Library of Congress Cataloging-in-Publication Data

Singh, Narinder, 1956-

An artificial intelligence approach to test generation.

(The Kluwer international series in engineering and computer science ; 19)
Bibliography: p.
1. Integrated circuits—Very large scale integration —Testing—Data processing. 2. Expert systems (Computer science) 3. Artificial intelligence. I. Title.
II. Series.
TK7874.S533 1986 621.395 86-20892
ISBN 0-89838-185-1

Printed in the United States of America

Contents

List of Figures vii

List of Tables ix

Preface xi

1 Introduction **1**

1.1 Motivation 1

1.2 Exploiting Design Morphology 2

1.3 Methodology 4

1.4 Example 6

1.5 Relationship to other work 11

1.5.1 Reasoning about devices 11

1.5.2 Reformulating representations 13

1.6 Overview 15

2 Reformulation **19**

2.1 Devices and Designs 20

2.1.1 Definition of a Device 20

2.1.2 Definition of a Design 21

2.2 Reformulating Designs 24

2.2.1 Abstracting Designs 25

2.2.2 Repartitioning Designs 48

2.2.3 Making Design Knowledge Explicit/Implicit 49

2.3 Design Correctness 50
2.3.1 Correctness for an Abstraction Level 51
2.3.2 Correctness Between Adjacent Abstraction Levels 55
2.4 Automatically Reformulating Designs 60
2.5 Manually Reformulating Designs 64

3 General Representation and Reasoning **67**
3.1 Requirements for a Design Description Language 67
3.2 Syntax and Semantics for Predicate Calculus 69
3.3 Describing Designs 73
3.4 Automated Deduction 76
3.4.1 Simulation 81
3.4.2 Diagnosis 81
3.4.3 Test Generation 83
3.4.4 Control 85
3.5 Utility of General Representation and Reasoning 88
3.5.1 Advantages 88
3.5.2 Disadvantages 91

4 Test Generation **93**
4.1 Task Definition 94
4.2 Previous Work 97
4.3 The Saturn Test Generation System 102
4.3.1 Algorithm 103
4.3.2 Example 106
4.3.3 Control Strategies to Increase Efficiency 111
4.3.4 Empirical Evaluation 128

5 Conclusion **133**
5.1 Summary of Key Ideas 133
5.2 Further Work 135
5.3 Implementation State 136

A Printer Adapter Card **137**

B Tests for the Printer Adapter Card **165**

Bibliography **189**

List of Figures

1.1 A picture of a multiplier device with n bit inputs. 7
1.2 A flat gate-level design of the multiplier. 8
1.3 A high-level design of the multiplier. 10

2.1 Picture of the device D74. 21
2.2 The set of objects in a design of the device D74. 24
2.3 An example of structural abstraction for a full-adder. . . . 27
2.4 Search space for controlling the carry output for the gate-level design. 29
2.5 Search space for the abstracted full-adder. 32
2.6 An example of spatial abstraction for an adder. 36
2.7 An example of temporal abstraction for an adder. 39
2.8 An example of value abstraction for a multiplier. 42
2.9 A device made up of two multipliers. 43
2.10 Search space using the value abstracted behavior of multipliers. 44
2.11 Search space using the integer behavior of multipliers. . . 45
2.12 An example of functional abstraction for an adder. 46
2.13 Search space for I/O table behavior formulation. 47
2.14 Design specification and actual behavior of an inverter. . . 52
2.15 Homomorphism between a device and a design. 53
2.16 Relating the behavior of the inverter and its design. . . . 54
2.17 Gate level and abstracted design of the carry function. . . 56
2.18 Output for the gate level and abstract design formulation of the carry function. 58

2.19 Transitions due to reconvergent fanouts. 59

3.1 A picture of the structure of a design for the device D74. 74
3.2 The automated deduction process. 77

4.1 Control structure of the D-algorithm. 98
4.2 Control structure of Saturn for achieving tests. 105
4.3 A design for an adder with 2 bit inputs. 107
4.4 Controlling the output of a full-adder to 0. 117
4.5 Costs for controlling the ports of a full-adder. 124
4.6 Costs for observing the ports of a full-adder. 124
4.7 A picture of a design for an adder with 4 bit inputs. 125
4.8 A test for the *or* gate. 127

A.1 The IBM specification for the Printer Adapter card. 139
A.2 An abstract formulation of the Printer Adapter card. 140

List of Tables

1.1 Cost of test generation using a flat and high-level design formulation. 9

2.1 Behavior of an *or* gate and the carry output of the full-adder. 28
2.2 Behavior of the full-adders and the adder. 38
2.3 Behavior of a register and the adder. 40
2.4 Accurate and approximate specification of the carry function. 56

3.1 Syntax for Predicate Calculus. 70
3.2 Mappings for the Interpretation and Assignment functions. 71
3.3 Definition of the semantic function Φ for terms. 71
3.4 Semantics for predicate calculus. 72
3.5 Structural description for D74. 75
3.6 Behavior description for D74. 76
3.7 A trace of resolution for simulating a design. 82
3.8 A trace of resolution residue for diagnosing faults. 84
3.9 A trace of resolution residue for generating a test. 86

4.1 Uncompressed tests for a full-adder. 108
4.2 Compressed tests for a full-adder. 110
4.3 Compressed and abstracted tests for a full-adder. 110

4.4 Compressed, abstracted, and generalized tests for a full-adder. 111
4.5 Conditional value propagation rules for an *and* gate. . . . 113
4.6 Propagating unknown conditional values. 115
4.7 A sequence of goal reductions using search. 118
4.8 Impact of meta-level control on search. 126
4.9 Impact of caching tests for prototypes. 126
4.10 Impact of caching subgoal solutions for test generation. . 128
4.11 Utility of refining designs. 129
4.12 Advantage of abstracting a design description. 130

Preface

I am indebted to my thesis advisor, Michael Genesereth, for his guidance, inspiration, and support which has made this research possible. As a teacher and a sounding board for new ideas, Mike was extremely helpful in pointing out flaws, and suggesting new directions to explore.

I would also like to thank Harold Brown for introducing me to the application of artificial intelligence to reasoning about designs, and his many valuable comments as a reader of this thesis. Significant contributions by the other members of my reading committee, Mark Horowitz, and Allen Peterson have greatly improved the content and organization of this thesis by forcing me to communicate my ideas more clearly.

I am extremely grateful to the other members of the Logic Group at the Heuristic Programming Project for being a sounding board for my ideas, and providing useful comments. In particular, I would like to thank Matt Ginsberg, Vineet Singh, Devika Subramanian, Richard Trietel, Dave Smith, Jock Mackinlay, and Glenn Kramer for their pointed criticisms.

This research was supported by Schlumberger Palo Alto Research (previously Fairchild Laboratory for Artificial Intelligence). I am grateful to Peter Hart, the former head of the AI lab, and his successor Marty Tenenbaum for providing an excellent environment for performing this research.

Finally, I would like to thank my wife Sue for her support through the difficult and trying years as a graduate student, and my parents for encouraging me to pursue my education in the U.S.

AN ARTIFICIAL INTELLIGENCE APPROACH TO TEST GENERATION

Chapter 1

Introduction

1.1 Motivation

In this thesis we are concerned with managing the time and space efficiency of reasoning about complex devices, involving tasks that span their life-cycle from design to manufacture and maintenance in the field. An integral part of building and maintaining systems involves representing and reasoning with design descriptions for a collection of tasks. These include verifying the correctness of a design, simulating a design to predict the output of the device for a given set of inputs, generating tests to verify the correct operation of a device, and diagnosing a device with a failure. Each of these tasks can be viewed from the AI perspective as reasoning over a collection of facts in a knowledge base.

In order to reason about a device we must formulate a design of the device (defining the parts, their interconnection, behavior, etc.), encode this design in a representation language, and finally use an inference method to reason with this design description.

With the advances being made in the technology of manufacturing devices it is now possible to build digital systems of unprecedented complexity. The complexity of the devices has led directly to the complexity of representing extremely large design descriptions. In addition, the complexity of the devices has greatly increased the size of the search space, and therefore the time required for reasoning about these devices. This is especially true for tasks such as test generation where the cost is exponential in the depth of the circuit from the inputs to the outputs. As the physical geometries of the components in the integrated circuits of a system are reduced, the number of components per chip increases quadratically. However, the number of input/output pins at the perimeter of the chips only increases linearly. This increases the depth of the circuit from the inputs to the outputs (by reducing the controllability and observability), and thus dramatically increases the cost for tasks like test generation.

This thesis is a part of the overall space of representing and reasoning about devices. The main contribution of this thesis is in the study of the utility of high-level design descriptions for a collection of tasks within the same framework. We will present examples from the domain of digital circuits, though the methodology proposed in this thesis is equally applicable in other domains where it is possible to decompose a device into a collection of components with well defined behavior that interact with each other along specific communication channels. In this thesis we examine how a design may be reformulated (though the automation of this process has not been mechanized), and describe why certain formulations are better than others. In addition, we define a correctness criteria to ensure the different abstractions of a design formulation are consistent with each other. We demonstrate the utility of reasoning with high-level design formulations for the specific task of test generation in the SATURN system, though we have demonstrated the utility of this approach for other tasks such as simulation [52] and diagnosis [23].

1.2 Exploiting Design Morphology

In this thesis we propose that an effective way of managing the complexity of reasoning about devices is to capture the *morphology* of a device in its design. The morphology of a device corresponds to its subparts, the properties of these subparts, and the relationships between them. The subparts of a device can be examined along a collection of dimensions, e.g., its physical structure, its functional organization, etc. In addition, we can examine the subparts of a device at a collection of abstraction levels along these dimensions, e.g., chips, boards and racks for its physical structure, and gates, combinational blocks and registers for its functional organization. The properties of a device include the relationships between the parts at different abstraction levels. For example, for a functional partitioning, the morphology of the device includes the relationships between a port value at an abstract level and the corresponding port value(s) at the next lower level.

A key idea is to capture a design description at a collection of levels that are *related* to each other. That is, the design includes a specification of the device at a collection of levels *and* the specifications for translating information across the abstraction boundaries. In order to be useful, the different abstractions of a design formulation must be correct with respect to each other and the device. This correctness is based on defining a homomorphism between any two abstraction levels, and between the device and an abstraction level. The homomorphisms define the degree to which the different abstractions are related to each other, which can range from equivalence (no loss of information), to an approximation of the exact values and/or temporal behavior.

The purpose of capturing the morphology of the device in its design is to take advantage of the properties of the device and its subparts in reasoning about it. By exploiting these properties we can improve both the efficiency of the reasoning task and the quality of the solutions. Greater efficiency is made possible by reasoning at a more abstract level, where a single inference step corresponds to a large number of inferences at the lower abstraction levels. This has the impact of reducing both the depth of the search space, and reducing the branching factor of the nodes in the search space. In addition, in problems where there are multiple solutions to a task, with differing utility, we can exploit the organization of the design to reason at the appropriate level to improve the quality of the solutions. For example, when generating tests for a device, the quality of the solutions is inversely proportional to the length of the test vectors returned. We can minimize the length of the test vectors by starting the test generation process from the level at which the fault models are defined.

The relevant properties of a device that should be captured in its design can be task dependent. For example, for simulation the important properties are the values at the outputs of the device as a function of the inputs. For test generation the important properties include how to control the inputs of the device to test each possible failure, and the value to observe at the output of the device for this test. In addition, for diagnosis, it is important to capture the relation between a failure (a set of inputs and the erroneous outputs) and the set of parts whose failure would cause this overall device failure.

If the morphology of the device is not captured in its design it is important to reformulate the design in order to reason effectively about the device. It is possible to automate some of the reformulation operations for a given task, e.g., partitioning a design to define a structural or behavioral abstraction. The partitioning can be based on a schema based substitution of a given design fragment with a reformulated design fragment for a given task. For example, for simulation, one schema might define that a collection of full-adders connected in a ripple-carry fashion can be abstracted as an adder. In addition, the schema can specify that the output of the adder is the sum of its inputs. These substitutions can be based on matching symbolic expressions, or checking for subgraph isomorphisms between the design and the schema. Similarly, if the task is to partition a design into a collection of physical units (boards, racks, etc.), then we can automatically partition a design, based on its connectivity, into clusters that are strongly interconnected locally and only weakly interconnected with other clusters. We can define the behavior of each physical unit by composing the behaviors of the subparts of a unit along their interconnections.

In the absence of an automatic reformulator, a useful design formulation to consider initially for all tasks is the one created in the design process. Due to the complexity of large devices, it is impossible to design them in a single refinement process from the initial specifications. This is especially true since design is usually an incremental refinement process where both the specification and the design co-evolve [9]. Since the device is fabricated using its design, its morphology reflects the more abstract formulations of its design. The formulation captured in the design process at a collection of abstraction levels can be a useful tradeoff between a flat low-level design formulation, and the ideal design formulation for many tasks. The flat low-level formulation will be inefficient due to the extremely large number of components through which information must be propagated. Similarly, the size of the ideal design can be extremely large for a given task due to the large number of properties that must be recorded. For example, for test generation, the ideal formulation will specify how to control and observe the value of every internal port in a device.

1.3 Methodology

In order to capture the morphology of the device at arbitrary abstraction levels we use a device independent language for encoding design descriptions. This independence permits using the same language to describe a design at a collection of abstraction levels, e.g., gate level, register transfer level, instruction set level, and algorithmic level. The language permits the user to define the abstraction levels at which to specify a design by allowing the user to extend the vocabulary to define a set of prototypes and behavior primitives for each abstraction level. The language, called Corona, is based on predicate calculus, and is described further in [51].

The declarative nature of the design description language permits using the same description for a collection of tasks. For example, reasoning forward to determine the outputs of a device for some inputs (as in simulation), reasoning backwards to deduce the inputs that produce a given output (as in test generation and diagnosis), or a combination of these two— given some inputs and some outputs, constraining the other inputs and outputs. This generality is made possible by the absence of side effects in the language, and the invertability of the primitive predicates in the language.

To exploit the design at a collection of abstraction levels in the same framework we use a device independent reasoning method to achieve a task. For example, the same inference procedure is used to generate tests at the boolean level and the register transfer level. The reasoning method explored in this thesis is based on resolution which is a complete inference procedure for first order logic. The efficiency and the quality of the solutions are improved by employing a meta-level architecture to guide the problem solver to reason at the appropriate abstraction level, and to first choose the promising paths in the search space at this abstraction level. This meta-level control is built on top of the MRS knowledge representation system [49]. In addition, the efficiency of the problem solver is increased by employing a collection of control strategies, e.g., caching, consistency checking, and constraint propagation.

The generality of the representation and reasoning methods have a cost overhead associated with them. For small designs this generality will lead to an inefficiency. However, for large designs, this generality coupled with the high-level design descriptions can lead to combinatoric savings, and mean the difference between being able to solve a problem or not.

This methodology has been developed in the context of the Helios design consultation system [21] in which the morphology of a design is exploited for a collection of tasks, including simulation [52], test generation and diagnosis [23]. An important part of these tasks requires propagating values through the design, for example, propagating values from the inputs to the outputs for simulation, and propagating values from the outputs to the inputs and vice-versa for test generation and diagnosis.

In this thesis we will illustrate the utility of this methodology for test generation, although the utility of this methodology has also been demonstrated for simulation and diagnosis. Test generation is more interesting than simulation since it involves reasoning forwards and backwards through the components of a design. Reasoning backwards through a component can lead to an explosion in the size of the search space, since there can be many inputs that produce the same output. As a task, test generation is interesting since we must manage the complexity of its large search space effectively in order to generate tests. We are not using diagnosis to illustrate this methodology since it shares many subtasks in common with test generation, and its other subtasks do not contribute to any additional computational complexity.

1.4 Example

In this section we will present a simple example which illustrates the utility of reasoning with high level design formulations for test generation. We will first illustrate the test generation procedure using a flat low-level design formulation of a multiplier, and later show the advantages of using a higher-level design formulation that captures the morphology of the device. In this section we will only present a brief description of the test generation task, leaving the more detailed discussion to Chapter 4.

Test generation involves generating a set of tests that verify the correct operation of a physical device. Each test checks the behavior of a collection of components in the device, and specifies the values of some inputs and the expected outputs for these inputs. If the outputs of a device are the same as the expected outputs for the test inputs, then the subparts being tested are assumed to be functioning correctly (and vice-versa). The collection of tests generated for a device should exhaustively test the behavior of all the parts of a device that can fail.

The test generation algorithm examined here first selects a local test for a subpart, and then propagates this test to the inputs and outputs of the device. A local test for a subpart specifies the inputs and outputs for the subpart alone. For example, a local test for an *and* gate specifies that its output should be 0 when its first input is 0. In order to achieve this test the test generator must figure out how to control the inputs of the device so that the first input of the *and* gate is 0 and its output is observable at an output of the device.

Suppose we have to generate tests for a multiplier device, which is pictured in Figure 1.1. The multiplier has two inputs *in*1 and *in*2 and a single output *out*. Each input is n bits wide, and the output is $2n$ bits

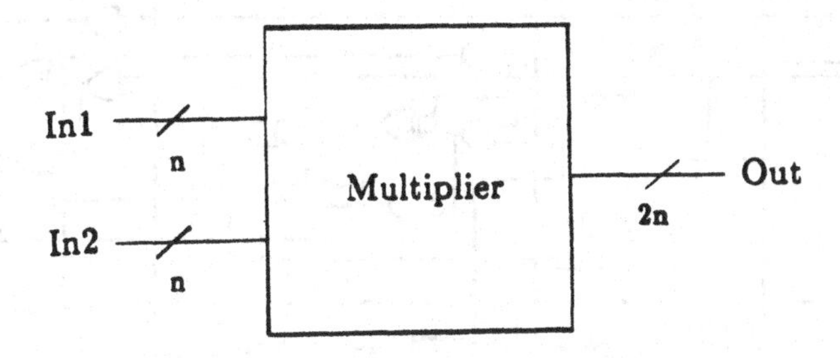

Figure 1.1: A picture of a multiplier device with n bit inputs.

wide. For this device assume that we have to test the behavior of each boolean gate in the device. In generating tests for this multiplier further assume that it is only possible to directly control the inputs *in*1 and *in*2, and it is only possible to directly observe the output *out*.

The ideal design formulation would specify how to control the directly controllable inputs to control the inputs of the subpart being tested, and it would also specify how to control the directly controllable inputs to propagate the output of the subpart being tested to a directly observable output. Unfortunately this information is usually not provided by a design, and each test must be achieved by propagating values through the adjacent subparts. In order to control the input of a subpart we must control the output of the subpart it is connected to. Similarly, in order to control the output of a subpart we must control some of its inputs. In order to observe the output of a subpart we must observe the input of the subpart it is connected to. Similarly, in order to observe the input of a subpart we must control some of the other inputs of this subpart and observe its output.

A flat gate-level formulation of a design for the multiplier is pictured in Figure 1.2.[1] The inputs of the multiplier are on the left side, and its outputs are on the right side. This figure highlights the propagation of values through the design for a test for an exclusive-or gate. The bold-faced connections show the other ports in the design whose values must be controlled/observed. The main disadvantage of the flat gate-level design formulation is the large number of components through which

[1] This figure is a part of the entire design.

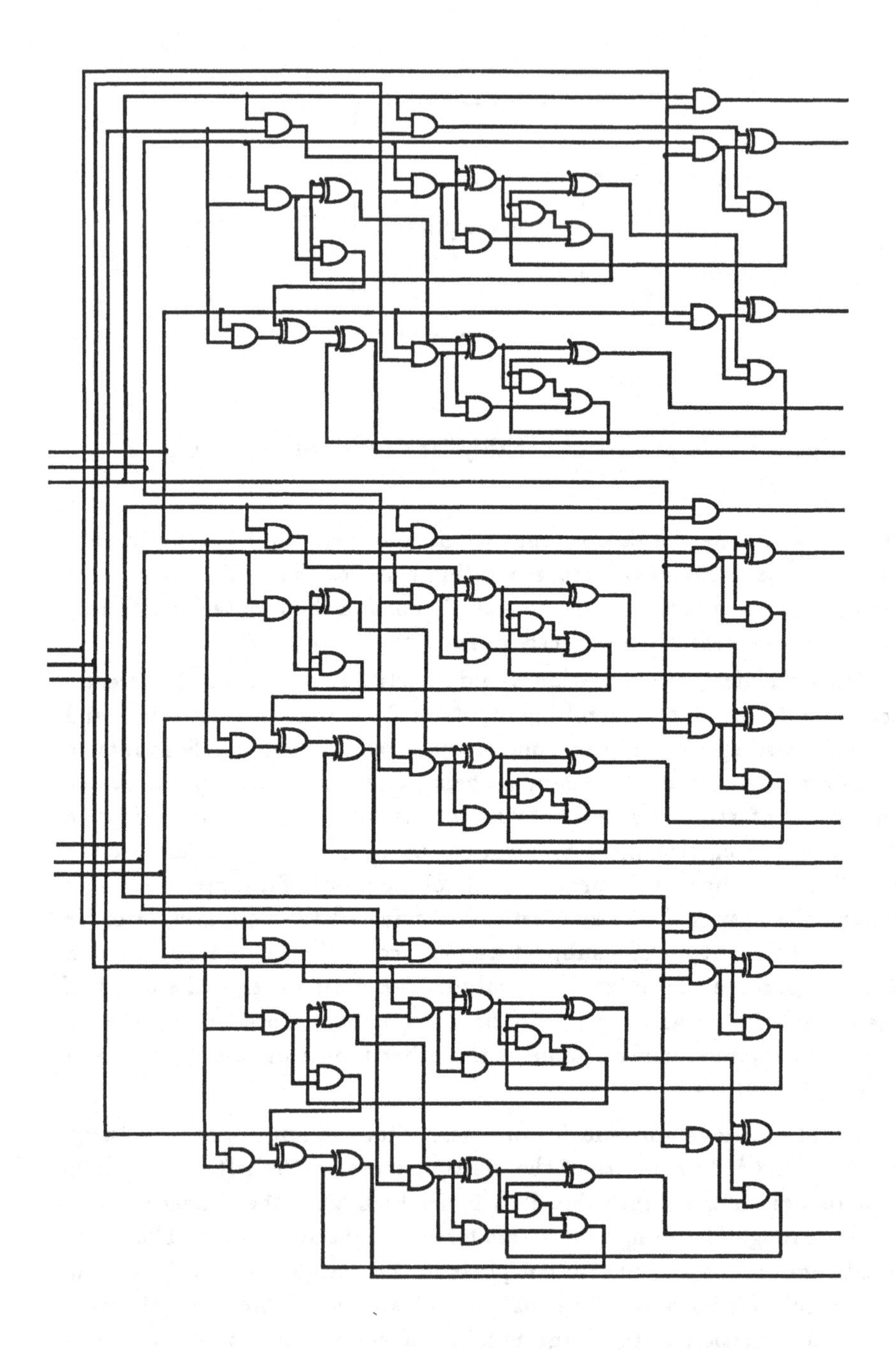

Figure 1.2: A flat gate-level design of the multiplier.

k=1.5	n=2	n=3	n=4	n=5
Flat	11	240	1.7×10^4	4×10^6
Reformulated	11	40	130	440

Table 1.1: Cost of test generation using a flat and high-level design formulation.

information must be propagated to achieve each test.

A higher-level design formulation of the multiplier is pictured in Figure 1.3. In this design the structure and behavior of the multiplier is decomposed into multiplier slices, bit multipliers, shifters, adders, and boolean gates. The test generator takes advantage of the reformulated design by always propagating information at the most abstract level possible. The propagation of values through the design for the same test for the exclusive-or gate is shown by the boldfaced connections in the figure. The test for the gate is first propagated to the boundary of its full-adder. The test is next abstracted to the higher level and propagated through the other full-adders of the adder, and so on to the inputs/outputs of the multiplier. At each hierarchy boundary the test is abstracted to the next higher level, and the propagation is continued at the higher level.

The main advantage of reasoning with the high-level design formulation is the reduction in the number of components through which information must be propagated. This is made possible by ignoring the subparts of a component when propagating information through it. For example, in order to control an output of a full-adder we only examine its high level behavior specification, and ignore its substructure.

In the worst case, the cost of test generation is exponential in the number of components through which information must be propagated. For the flat design, this number is approximately k^{n^2}, where n is the number of bits at each input of the multiplier, and k is the average number of ways to control/observe a port. For the higher-level design formulation, however, this number is approximately k^n. Table 1.1 compares the cost of test generation for these two design formulations for different size multipliers (we have chosen a typical value of 1.5 for k). The ratio of the cost of the flat and high-level design formulations grows exponentially in the size of the multiplier. Consequently, by reformulating designs we can dramatically reduce the cost of test generation and extend the complexity of the devices we can handle.

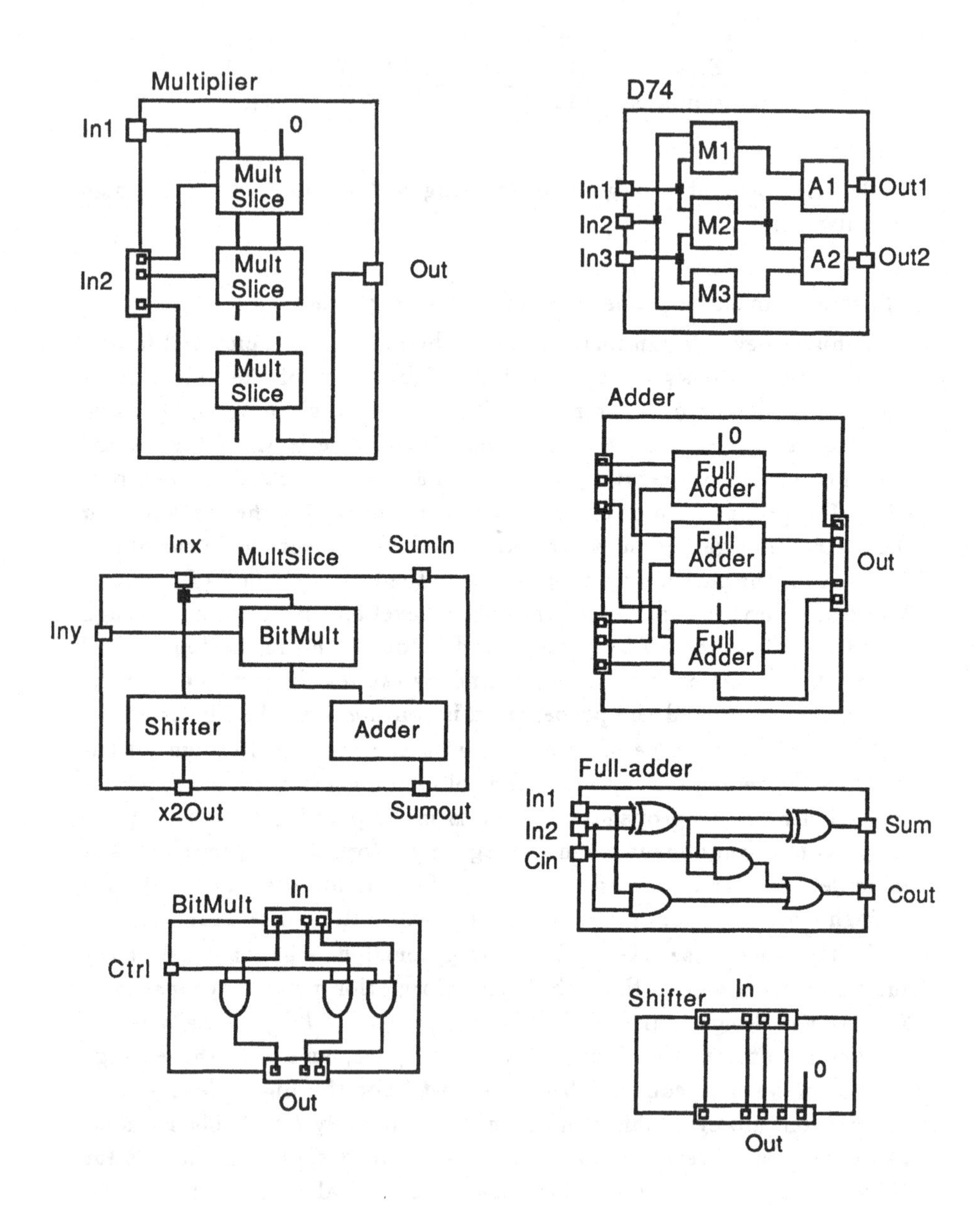

Figure 1.3: A high-level design of the multiplier.

1.5 Relationship to other work

The key idea proposed in this thesis is that the morphology of the device should be captured in its design to increase both the efficiency and the quality of solutions for a collection of tasks for the device, and also to reduce the size of the design. In this section we will compare this approach with other work related to reasoning about devices, and other research in AI illustrating the utility of reasoning with the appropriate knowledge formulation to increase the efficiency for a given task.

1.5.1 Reasoning about devices

Earlier approaches to reasoning about devices have forced the specification of designs at specific abstraction levels. The different design description languages are useful intermediate islands in the process of refining an abstract design specification into an implementation. Each language addresses a different set of concerns appropriate at its abstraction level, and provides a set of primitives and rules for combining these into composite design descriptions. For example, architecture description languages like ISP [3] are used to define the appropriate instruction set and architecture for processors. The composition rules attempt to avoid the typical errors at a given abstraction level, for example, register transfer languages permit detecting errors due to incorrect clocking schemes. Similarly, switch level languages, such as the one presented in [10], permit detecting incorrect designs where signal values violate digital behavior. Layout languages like SILT [14] permit detecting errors due to the physical proximity of components that might lead to a short (or other incorrect behavior) during the fabrication process. These issues are further addressed in [56].

Unfortunately these different design languages have not been integrated in a single environment. As a result, a design must be manually translated between the different languages, and it is impossible to check the consistency between the different descriptions.

Recently there has been considerable activity in the use of hierarchical languages for specifying designs, e.g., ADLIB [27] DPL [6] and innumerable others which can be found in any recent Design Automation Conference proceedings. These languages permit specifying a design at a collection of abstraction levels in a single framework. For example, in ADLIB it is possible to specify the structure and behavior of a design hierarchically, where the behavior is specified in a language that is a superset of Pascal. Although it is possible to capture the morphology of the device in these languages, the types of reasoning possible with these descriptions is quite limited. For example, the procedural representation of design knowledge in ADLIB is best suited for simulation, and not other tasks such as test generation and verification. This is due to the fact that the procedural representations are hard to invert and combine with each other. This is especially true if the language permits side-effects and special constructs for specific tasks, e.g., the *waitfor* construct in ADLIB which can have side-effects due to its dependency on global variables. Other language constructs, such as *sensitize*, appear to exist only to guide the scheduler for simulating the design. Similarly, DPL design descriptions can only be used for generating mask layout for fabricating a device.

More interesting reasoning tasks, such as test generation, are usually performed at restricted abstraction levels. For example, the test generation algorithm proposed by Roth [48] is only defined for designs at the boolean level. Similarly, the implicit enumeration test generation algorithm PODEM [26] is also defined at the boolean level. Other approaches have integrated a few languages in a single environment, e.g., the SCIRTSS [28] test generation system which integrates boolean design descriptions for combinational circuits with a finite-state-machine model for the transfer of information between the registers of the machine. An exception to this is the HITEST test generation system [47] which attempts to generate tests for a hierarchical design description using the PODEM algorithm, with guidance from the user when the system runs into trouble (after more than 9 backtracking operations). The system attempts to capture the expert knowledge of the test engineer by allowing him to specify how to test design fragments, e.g., fragments that would be too difficult for HITEST to generate tests for automatically.

The design verification system VERIFY [5] is another exception, where a hierarchical design description is used to formally prove the correctness of a design. At each hierarchy boundary the system compares the symbolic specification of a module with the composition of the symbolic specifications of its subparts, based on their interconnection. The equivalence between the two symbolic expressions is proved automatically as far as possible, with the user suggesting different simplification operations in case of failure. At present, the system is restricted to proving the functionality of the design without taking into account its temporal behavior. The main advantage of a hierarchical design for this task is in identifying identical components, so that the verification operations need not be repeated. In addition, it is more efficient to prove the equivalence between many pairs of simpler expressions than between a single pair of complex expressions.

1.5.2 Reformulating representations

The idea that the formulation of knowledge can have an impact on the efficiency of the reasoning process is hardly a new one to AI. Researchers have carved out different subproblems from the vast area of reformulation, examining its utility in different problem domains. However, there has been limited success in developing a general theory of reformulation that might be applicable to a broad class of problems. In general, reformulation involves changing the objects and/or the inference mechanism to increase the efficiency of the problem solving activity, or to capture more abstract relations between objects. In this section we will only examine a few important efforts in this area, leaving many others unexplored.

One of the eariest examples illustrating the utility of selecting the appropriate knowledge formulation is the mutilated checkerboard problem presented by McCarthy [37]. In this problem, two diagonally opposite corners of an eight-by-eight checkerboard are cut out, and the problem is to figure out if this board can be covered exactly with dominos. The search space is extremely large if each square is identified by its coordinates, and each domino placement records the two squares which it covers. A problem solver using this design will exhaustively search this space, and find that there is no solution. Alternatively, if we only take into account the color of each square (black or white), and the difference between the number of uncovered black squares and uncovered white squares, then it is simple to prove that the problem is unsolvable. Initially there are two more uncovered black squares than white squares, and this difference is unchanged by any domino placements. Consequently, when all the white squares are covered there must still be two uncovered black squares, and it is impossible to cover two black squares with one domino.

Unfortunately, this example was only meant to demonstrate the difficulty that a problem solver would face when given the original problem formulation. The reformulation operations were performed manually, and no constructive procedure was presented for doing this automatically.

Another example illustrating the utility of reasoning with reformated descriptions is by Amarel in his 1968 paper on the missionaries and cannibals problem [2]. This problem involves finding the simplest schedule to transfer three missionaries and three cannibals from one bank of a river to the other. A boat is available that can only carry one or two people. At no place should the number of cannibals exceed the number of missionaries. After a series of representation transformations Amarel shows that the problem can be solved easily. These transformations focus on macro-operator formation (replacing a collection of actions with a single action), detecting temporal symmetry in the problem, and detecting symmetries for larger problems (more than three missionaries and cannibals). Again, these transformations are pencil and paper exercises and they are not justified, i.e., why certain transformations were performed, how they were discovered, why they are correct, etc. While Amarel comments on possible mechanizations of these transformations, these seem overly optimistic, even with the current AI technology.

More recently, Richard Korf has characterized all representation shifts along the two dimensions of *information structure* and *information quantity* [33]. The former corresponds to isomorphic transformations that change the information structure without changing the quantity. The latter corresponds to homomorphic transformations that change (reduce) the information quantity without changing the structure. In this paradigm, reformulations can be viewed as heuristic search in a space of possible representations, and an integral part of problem solving is viewed as transforming representations till the problem is trivial to solve. This paradigm is illustrated for the following problems: tic-tac-toe, integer arithmetic, tower of Hanoi, the arrow puzzle, five puzzle, mutilated checkerboard, and floor plan design. The paper presents a language for specifying representations and transformation schemas. Primitive transformations can be automatically inverted and combined to define more powerful transformations. Unfortunately, the transformations to be applied must be selected manually. In addition, there is no notion of correctness for these transformations.

Up to this point we have examined representation transformations that are motivated by increasing the efficiency of the problem solver. On the other hand, for classification problems the task *is* to transform a low level representation of a collection of objects into a more abstract representation.

For example, the circuit understanding system Qual [15] automatically formulates a hierarchical representation of how a circuit achieves its ultimate purpose. The system constructs a mechanization graph of the circuit to define fragments which can be used to define a hierarchical description of the circuit behavior. The recognition proceeds from the bottom up, where larger fragments are constructed from smaller ones using *abstraction* rules. A typical abstraction rule, for example, ignores fragments of passive elements in a design. The recognition process proceeds until a single fragment remains. The system can recognize a wide class of amplifiers and power supplies. However, the system cannot analyze the quiescent behavior of a device, nor can it handle logic (digital) circuits.

An important subtask in the circuit verification system VERIFY [5] is to compose the behavior of the subparts of a module to derive a specification of the behavior for the module as a whole. The original design is specified hierarchically, and the goal of this system is to compare the specified behavior of a module with the behavior generated by examining its subparts. The circuit partitioning task is avoided since the original design is specified hierarchically. This system is applicable to a wide class of problems for components with discrete behavior, e.g., digital circuits.

Another example from the software domain is the REASON system [50] for understanding programs. The goal of the system is to help in the evolution of program design by understanding the impact of changes on the program. The system matches sections of the program with a library of pre-analyzed plan fragments. The different plan fragments are connected together by the control structure of the program. This network of plan fragments provides a teleological structure of the program, which defines how different parts of the program interact with each other to achieve the desired overall behavior.

This section has briefly illustrated some of the related work on reasoning about devices, and reformulating representations. Each of these systems has carved out a part of the overall problem by restricting the domain, and/or the reasoning tasks.

1.6 Overview

This section will present a brief overview of the remainder of this thesis. There are three important parts of this thesis. The first part is concerned with reformulating designs, while the second part describes the utility of general methods for representing and reasoning about devices. Finally, the last part describes test generation in detail. Each of these parts is described in a separate chapter which can be read independently.

In this chapter we have informally discussed the advantages of reasoning with high level design formulations. In Chapter 2 we will present a more formal discussion of reformulation. We will first present a definition of a device, which is a physical entity, and a design, which is a specification of this entity. We will next examine the different dimensions along which a design can be reformulated, and demonstrate the utility of performing these reformulations by examining the size of the search space for different design formulations. Chapter 2 will show that the different reformulation operations can reduce both the depth and the branching factor of the search space. In addition, it will show that it is possible to avoid search by reformulating designs, which permits solving problems using constraint propagation. This chapter will show that the different reformulation operations that reduce the size of the search space have the additional benefit of reducing the size of the design.

In order to take advantage of the higher-level design formulations we must ensure that these formulations are a correct specification of the device. In Chapter 2 we will also describe the criteria for a design to be a correct specification of a device. This correctness criteria is extended to ensure that the specifications of a design at different abstract levels are mutually consistent.

Lastly, Chapter 2 will suggest possible ways of automatically reformulating designs. In the absence of an automatic design reformulation system all is not lost. We will show that certain reformulation operations cannot be automated, and that others can be performed better manually.

Chapter 3 describes the use of general methods based on logic for representing and reasoning about designs. This chapter will define the requirements for a design description language, and present the syntax and semantics of a language based on predicate calculus to describe designs. In addition, the chapter will show how to represent the structure and behavior of a design using this language. In addition, this chapter will define a general inference procedure for reasoning about designs, and show how it is possible to use this inference procedure to simulate, diagnose, and test devices. The last section of Chapter 3 will describe the utility of using general methods for representing and reasoning about designs.

Chapter 4 examines the test generation task in greater detail. The first part of this chapter defines the test generation task, and reviews the previous work in this area and describes their limitations. The second part describes the Saturn test generation system which uses the general representation and reasoning methods described in Chapter 3 to take advantage of the higher level design formulations in generating tests for a device. This discussion will ignore the details of the representation and reasoning methods, and concentrate on the control issues for efficiently managing the complexity of test generation, e.g., constraint propagation, consistency checking, caching tests and solutions to subgoals, and heuristics to guide search. Finally, the last section of Chapter 4 will present empirical results for the Saturn test generation system that validate the utility of reasoning with high level design formulations.

Chapter 5 will present a summary of the main ideas of this thesis, and suggests directions in which this work can be extended. It will also describe the current implementation state of the Saturn test generation system.

Those readers interested in a complete specification for a *real* design can read Appendix A, which presents an abstracted design formulation for the IBM PC printer adapter card in the language Corona [51], which is a prefixed form of predicate calculus with a Lisp-like syntax. Finally, Appendix B presents the tests that are generated by the Saturn test generation using the design formulation presented in Appendix A.

Chapter 2

Reformulation

The previous chapter described the difficulty in reasoning about complex devices, and outlined our approach to manage this complexity, i.e., reasoning with high level design formulations that capture the morphology of the device. In this chapter we will present a more formal discussion of reformulation. Before we define how designs can be reformulated, we present a precise definition of a device which is a physical entity, and a design, which is a specification of this device. We will next examine the different types of design reformulation operations, and demonstrate the utility of performing these reformulations by examining their impact on the size of the search space, and the size of the design. We will show that the different reformulation operations can reduce both the depth and branching factor of the search space, and also reduce the size of the design. In addition, we will show that it is possible to avoid search by reformulating designs and solving tasks by propagating symbolic constraints.

In order to be useful, the specification of a device at the different abstraction levels in the design must be a correct specification of the device. We will next define the criteria for a design to be a correct specification of a device. This is based on defining a mapping between the behavior of the device and the behavior specified by the design at each abstraction level. The correctness criteria is extended to ensure that the design specification at the different abstraction levels are consistent with each other.

The last part of this chapter will suggest possible mechanizations for some of these reformulation operations, and describe why certain reformulations cannot be automated. In the absence of an automatic reformulation system we will show that it is possible to reformulate designs manually, and that a useful design formulation to consider initially for all tasks is the one that is created in the design process.

2.1 Devices and Designs

In order to reason about a device we must conceptualize a formulation of the device as a design, encode this design in a representation language to form a design description, and apply an inference procedure over this description. In this section we will define and distinguish between a *device*, and a *design*.

2.1.1 Definition of a Device

A *device* is a specific arrangement of physical components. As a physical artifact, we can examine it along a collection of dimensions. In this thesis we will be concerned with examining the structure and behavior of a device, although it can also be examined along other dimensions, e.g., its manufacturing process, cost, etc.

The structure of a device is defined by the environment-independent relationships that hold between the physical entities that comprise it, e.g., the subparts of a device, their dimensions and relative positions. Similarly, the behavior of a device is defined by the environment dependent relations that exist between the physical entities that comprise it, e.g., the voltage at one point in a device as a function of the voltages at a collection of other points in the device.

An example of a device is given in Figure 2.1. This figure is a picture of a device named D74[1]. This device has three 2 bit inputs *a*, *b* and *c*, and two inputs for power and ground. In addition, the device has two 5 bit outputs *d* and *e*.

The structure of this device includes the eighteen input and output pins, the plastic casing, the wires from the pins to the integrated circuit inside the casing, and the structure of the integrated circuit itself. The behavior of this device is defined by the temporal relations that exist between its parts. For some collection of parts no such relationship exists, for example, between any two points on the plastic casing, or between any of the input pins. For other collection of parts, however,

[1]This device is not manufactured commercially.

Figure 2.1: Picture of the device D74.

such a relationship does exist. For example, if the voltage on the first two pins is 0 volts and the chip is powered (pin nine is 0 volts, and pin eighteen is 5 volts), the voltage on pins seven through seventeen is also 0 volts[2]. In fact there are many such relationships between the parts of the device, for example, between a collection of parts inside the integrated circuit. All temporal relationships between parts of a device are included in its behavior.

2.1.2 Definition of a Design

A *design* is a specification of a device. This specification is a formulation of a device at the knowledge level, independent of any symbols. A given design formulation reflects how we choose to view a device. For example, we could choose to view a device as a single function (a black box with a set of inputs and an output), or we could choose to view it as a composition of functions (a collection of interconnected black boxes), or a combination of the two. The formulation we choose will, in general, be incomplete, ignoring detail that is irrelevant to the tasks at hand.

A given design D is a three tuple $< O, F, R >$, where O is a set of objects, F is a set of functions and R is a set of relations. For every function in the formulation of a device there is a corresponding set of objects with which this function is associated (mapping from the arguments and result of the function to objects $\in O$). In addition, the elements in the domain and range of the functions are members of the set of objects O.

[2]The pins of the device are numbered one through eighteen, counter-clockwise, starting with the bottom left pin.

The relationships between various objects is defined by the elements of R.

The set of objects includes the modules, ports, connections and state-variables of the device, and the values used to describe the behavior of this device. Modules define the components of a design. Each module has a set of input and/or output ports which are the only points through which it can communicate with its environment. Communication between modules is defined by connections which relate the values of the ports at its two endpoints. State variables are used to define a partial history of the values at ports, or the internal state of modules. Objects of type *value* are elements of the domain/range of the functions that define the behavior for a device. Modules, ports and connections can be composite. The subparts of a module are its submodules and their connections. The subparts of a port are its subports, and the subpart of a connection is its submodule. These objects can be decomposed recursively down to a primitive set of modules, ports, and connections.

Every module, port, connection and state-variable $\in O$ corresponds to some n-tuple of three-dimensional physical regions in the device. For example, a given module in a design can correspond to a collection of integrated circuits in the device. Similarly, a port in a design can correspond to a collection of physical regions in the device. This correspondence defines an interpretation of the values (voltages) for the regions in the device and the value of the port.

Elements of F define functions between members of the set of objects O. These functions define the behavior for the modules, ports and connections. In describing behavior, the arguments of a function are the values of some ports/state-variables and a specification of the time at which each value is true (e.g., a single point in time, or a time interval between two points). The result of the function is a pair. The first element of the pair is the value of a port/state-variable, and the second element is a specification of the time at which this value is true.

The behavior of a module defines the temporal relations between the values of its ports and state variables. For a module with multiple outputs and state-variables, a separate function is defined for each output and state-variable. For a composite module we have a set of functions (one for each output and state-variable) defining its behavior at the high level, and a composition of functions defining it at the next lower level (and so recursively down to a primitive set of functions). Similarly, the behavior of a connection specifies the temporal relationships between the values of the two ports at its endpoints. The behavior of a port specifies the temporal relations between its value and the values of its subports.

In addition to defining the behavior of components, it is possible to define functional relationships between a collection of ports, all of which are not associated with the same module. Examples of this can be found in Chapter 4 where functional relationships between the inputs/outputs of a device and its internal ports are cached.

Members of R define the type of an object in O, and the relationships between these objects, e.g., the relations: module, port, connection, state-variable, submodule, subport, subconnection, and connected. The first four relations define the type of an object to be a module, port, connection or state-variable. A submodule relation defines a submodule of a module, and a subport relation defines a subport of a port. Similarly, a subconnection relation defines *the* submodule of a connection. This submodule has two ports, each corresponding to one end of the connection. The structure and behavior of the high level connection object is refined by this submodule, which itself may be further decomposed in terms of other modules, ports and connections. Finally, the connected relation defines the two ports at the endpoints of a connection object.

In general, there are many ways to formulate a design for a device. For example, one possible formulation for the device D74 is to conceptualize it as a black box, where the substructure is not explicated. This design corresponds to the 3-tuple $< O_1, F_1, R_1 >$. The set of objects O_1 consists of the objects labelled *d*74, *a*, *b*, *c*, *d* and *e* in Figure 2.2., and the set of real numbers. The object labelled *d*74 corresponds to the entire device, and the object labelled *a* corresponds to the first two pins of the device. The other objects *b*, *c*, *d* and *e* correspond to a collection of other pins of the device as pictured in Figure 2.2. This formulation does not define any connection or state-variable objects.

The set of functions F_1 consists of two functions. The first function defines the temporal relations between the values of the objects labelled *a*, *b* and *c*, and the value of the object labelled *d*. Similarly, the second function defines the temporal relations between the values of the objects labelled *a*, *b* and *c*, and the value of the object labelled *e*. The arguments of these functions are the values of the objects (each an integer), and the time at which these values are true (a real number). The result of each function is a pair, where the first element of the pair is the value of the result object (an integer), and the second element is time when this value is true (a real). The first element of the result of the first function defines the following relation between the values of the objects $d = a \times b + a \times c$. Similarly, the first element of the result of the second function defines

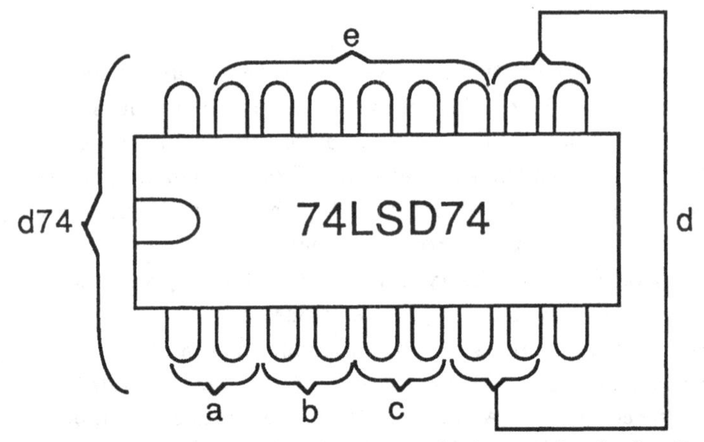

Figure 2.2: The set of objects in a design of the device D74.

the following relation between the values of the objects $e = b \times c + a \times c$. The second element of the result of both functions is equal to their time argument plus a number between 100 and 150 (depending on the other argument values, thus defining the delay of the device for different inputs).

The set R_1 consists of six relations. The first relation defines the object labelled *d74* in Figure 2.2 to be a module. Similarly, the remaining relations define each of the objects labelled a, b, c, d and e to be a port.

2.2 Reformulating Designs

The purpose of reformulating a design is to improve the efficiency and/or quality of solutions in reasoning about a device. In this section we will define the different types of reformulation operations, and examine the utility of these reformulations. Depending on the task, a combination of these reformulation operations can be applied to a design to transform it closer to the ideal design formulation.

Reformulation involves translating one design $D =< O, F, R >$ into another design $D' =< O', F', R' >$. The reformulated design D' must be a correct specification of the same device specified by D. By reformulating a design we are changing the way we wish to view a device. That is, we can choose to view a device as being composed of a different set of objects, and we can choose to view new functions/relations between these objects.

In reformulating a design $D =< O, F, R >$ into another design $D' =< O', F', R' >$ we can either: *abstract/refine* the existing design, choose a different partitioning for the design, or make explicit/implicit the functions and relations between the objects. A brief description of each of these reformulation operations is given below.

Abstraction corresponds to creating new objects in the design, each of which corresponds to a collection of existing objects in the original design. In addition, we can define new functions (e.g., behavior) and relations (e.g., connections between objects) for the newly created objects. In abstracting a design we do not throw away any information—the objects that are the subparts of the newly created objects are still a part of the design. Refining a design is the inverse of abstraction, and involves creating a collection of objects that refine an existing object in the original design. For example, we can refine a design by defining the substructure of an existing primitive object. Due to the duality of abstraction and refinement, the discussion in the following subsection for abstraction is equally applicable for refinement.

Repartitioning a design involves choosing a different set of objects for a design such that the primitive objects in the new and old design are the same. Finally, reformulating a design to make knowledge explicit/implicit selects a different space time tradeoff. All facts that could be deduced in the old design can still be deduced in the new design, and vice versa. However, some facts can be deduced more/less efficiently with the new design.

The remainder of this section will describe each of these reformulation operations in greater detail, and examine their utility.

2.2.1 Abstracting Designs

Abstraction corresponds to creating new objects in a design, and defining functional and relational properties for these objects. Each new object corresponds to a collection of existing objects in the original design. The new functions can define the behavior of the newly created objects, and specify the translation of information across the newly created abstraction boundaries. In abstracting a design we augment the information in the original design. The subparts of the newly created objects, and their functional and relational properties are still a part of the new design.

Formally, abstraction involves transforming a design $D = <O, F, R>$ into the design $D' = <O', F', R'>$ such that:

$$O \subseteq O' \wedge F \subseteq F' \wedge R \subseteq R'$$

Each newly created module $o'_i \in O' \setminus O$[3] corresponds to a collection of interconnected modules $o_{i,1}, \ldots, o_{i,k} \in O$. This abstraction defines a *new* partitioning on top of the existing partitioning of the original design D. The new functions in $F' \setminus F$ correspond to the functional relationships between the ports of the newly created modules, and between these ports and the ports of the substructure of the new modules. Similarly, the new relations in $R' \setminus R$ correspond to the relations between the newly created modules, and between these modules and their substructure.

There are five types of abstraction operations: structural, spatial, temporal, value, and functional. Structural abstraction corresponds to the case where there is a one-to-one mapping between a value at the abstract level and a value at the substructure. Similarly, spatial abstraction corresponds to the case where a value at the abstract level corresponds to a collection of values at the lower level, each at a different point in space, and temporal abstraction corresponds to the case where a single value at the abstract level corresponds to a collection of values at the lower level, each at a different point in time. Value abstraction corresponds to the case where the set of values at the lower level are partitioned into equivalence classes such that all values in the same equivalence class at the lower level map to a unique value at the abstract level. Finally, functional abstraction corresponds to the case where we reify a new function (give it a name a define its properties), and define the behavior of the new objects in terms of this function.

2.2.1.1 Structural Abstraction

In defining a structural abstraction the vocabulary for describing the behavior of a newly created module is the same as the vocabulary for describing the behavior of its subparts, i.e., there is a one-to-one mapping between a value at the abstract level and a value at the lower level.

An example of structural abstraction is given in Figure 2.3 for a full-adder device. The original design is defined in terms of a collection of boolean gates, and the new design aggregates this structure to define a new full-adder and its properties. The set of objects in the original design O are the 5 gates, the ports of these gates, the 12 connections, and the boolean truth values 0 (false) and 1 (true). The functions in the original design F define the behavior of each gate and connection. Similarly the relations in the original design R define the type of each component, and define the endpoints of the connections.

[3] $O' \setminus O$ means the set O' minus the set O.

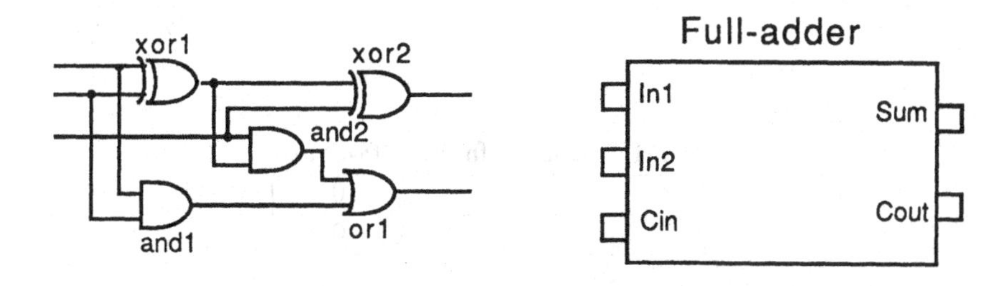

Figure 2.3: An example of structural abstraction for a full-adder.

The set of objects in the structurally abstracted design include the existing objects of the old design, and the new objects corresponding to the full-adder and its ports. The new design has one additional function for the behavior of the full-adder, and has 6 additional relations that define the type of the full-adder and its ports. A formal description of the new design is given below:

$$\begin{aligned} O' = O \cup & \{fa, fa_{in1}, fa_{in2}, fa_{cin}, fa_{sum}, fa_{cout}\} \\ F' = F \cup & \{behav_{fa}\} \\ R' = R \cup & \{module(fa), \\ & port(fa_{in1}, fa), port(fa_{in2}, fa), port(fa_{cin}, fa), \\ & port(fa_{sum}, fa), port(fa_{cout}, fa)\} \end{aligned}$$

The function $behav_{fa}$ denotes the behavior of the full-adder. The relation $module(fa)$ defines the object fa to be a module, and similarly the relation $port(fa_{cin}, fa)$ defines the object fa_{cin} to be a port of the module fa.

The behavior of the gates, connections, and the full-adder define the input/output relationships that exist between the values at their ports. For example, Table 2.1 describes the behavior of the *or* gate, the connections, and the carry output of the full-adder. We will describe the behavior of components using tables for clarity of presentation, although we are not making use of any special properties of tables. In the next chapter we will go into the details of actually representing these behav-

or1-in1	or1-in2	or1-out
0	0	0
x	1	1
1	x	1

Or

in	out
0	0
1	1

Connections

fa-in1	fa-in2	fa-cin	fa-cout
x	0	0	0
0	x	0	0
0	0	x	0
x	1	1	1
1	x	1	1
1	1	x	1

Carry output

Table 2.1: Behavior of an *or* gate and the carry output of the full-adder.

iors.

The entries in the table labelled "x" correspond to don't-care values, i.e., they can have any value. The left table defines the behavior of the *or* gate, and the behavior of the other gates is specified similarly. The middle table defines the behavior for all connections, i.e., this behavior specification is shared by all the connections. Finally, the right table defines the behavior of the *carry* output of the full-adder, and the behavior of the *sum* output can be defined similarly.

The remainder of this subsection will present an example illustrating the utility of reasoning with the reformulated design of the full-adder device, and later analyze the utility of structural abstraction in general.

Suppose we have the goal of controlling the carry output of the device to 1, or 0. The search space using the original low-level design formulation is given in Figure 2.4. The top-half of the figure shows the search space for controlling the carry output to 1, and the bottom half shows the search space for controlling the carry output to 0. In this figure we have used the name of a port to represent the goal of controlling its value to 1, and name of a port with an overbar to represent the goal of controlling its value to 0. In addition, adjacent port names stand for a conjunction of subgoals. For example, $a\bar{b}$ represents the goal of controlling port a to 1 and controlling port b to 0.

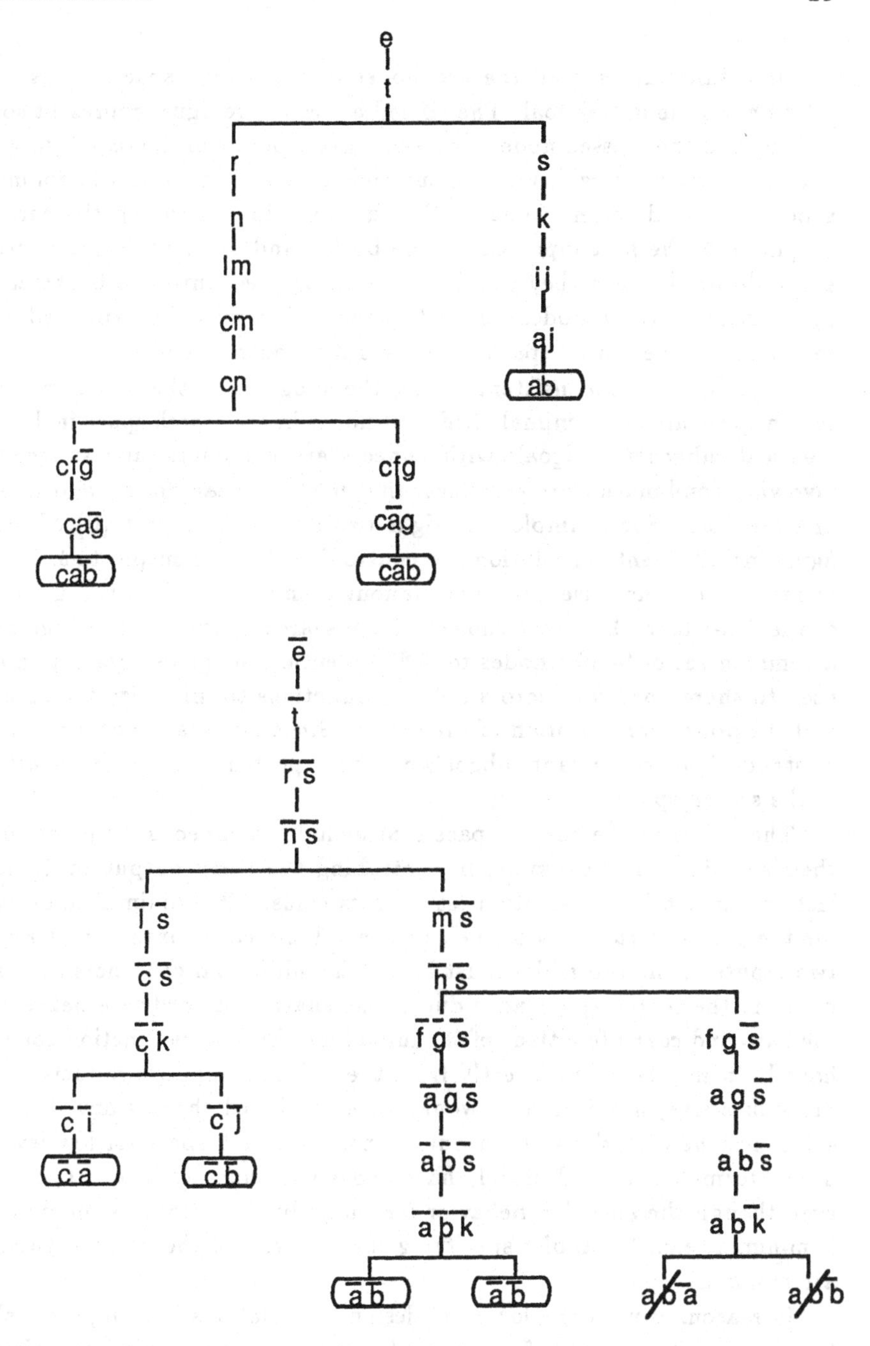

Figure 2.4: Search space for controlling the carry output for the gate-level design.

Unfortunately, not all the leaf nodes in the search space represent solutions to the initial goal. The circled nodes in the figure represent solutions, and the crossed nodes represent failure points in the search space due to inconsistencies (when the subgoals of a node require controlling a port to two different values). For example, in controlling the carry output to 0, the first input cannot be both 1 and 0. Inconsistent nodes are undesirable since they require a reasoning mechanism to backtrack upon reaching these nodes, and additional effort must be expended at each node in the search space to check if it is inconsistent.

In addition to inconsistent nodes, the subgoals at the nodes in the search space are not minimal. Different nodes in the search space include identical subgoals, subgoals with unnecessary conjuncts, and subgoals involving combinations of conjuncts that may be meaningless, and thus unachievable. For example, the right two solutions in part (b) of the figure include identical solutions for controlling the sum output to 0. The identical solutions arise due to the fanout points at the first two inputs of the full-adder. Different choices in the search space can converge at a common set of fanout nodes to define identical subgoals. Fanouts are used to share hardware across different functions to minimize the area, and the power consumption of the device. Since fanouts are common in digital designs, redundant subgoals can be expected to occur frequently in the search space.

The nodes in the search space also include extraneous conjuncts in their subgoals. For example, in controlling the carry output to 1 the last conjunct $\bar{b}$ in the solution $ca\bar{b}$ is extraneous. The minimal solution for the goal e is $ab \vee ac \vee bc$, i.e., to control the carry output to 1 any two inputs of the full-adder must be 1. The additional conjuncts at the nodes in the search space arise due to the sharing of hardware between the *sum* and *carry* functions of the full-adder. The *carry* function could have been implemented directly from the minimal sum-of-products expression above, however, this would have precluded sharing any of the sub-functions with the *sum* output. Thus, the search space for low-level design formulations will usually have nodes with non-minimal subgoals, even though the specified behavior for the individual low-level modules is minimal (e.g., the tables specifying the behavior of the boolean gates are minimal).

In reasoning with the low-level design formulations it is impractical to minimize the subgoals for each node, since this can require exploring the entire search space and performing an exponential minimization algorithm over this search space. For example, it is impractical to return the minimal solutions to the original goal because the minimization process requires knowing, and thus finding, all solutions. If we are interested in a single solution, as is the case for test generation, finding all solutions is clearly impractical. Even if all the solutions were known, minimization would be impractical since this problem is known to be NP-hard [22] (e.g., boolean minimization).

Having non-minimal subgoals at a node in the search space has a dramatic impact on the efficiency and completeness of the reasoning procedure. Nodes with identical subgoals use up resources without providing any novel information. Once a node with certain subgoals has been explored, all other nodes in the search space with identical subgoals are similar to inconsistent nodes. In addition, nodes with combinations of conjuncts that violate the intention of the designer may not have any solutions in the subtree below them in the search space (an example of this will be given later). Thus, exploring these nodes represents wasted effort. Also, these nodes are more detrimental than duplicate nodes since the inconsistencies will only be found after additional effort.

A more important problem, however, is in having extraneous conjuncts in the subgoals of a node. Let the minimal subgoal list for a node be g, and assume that the actual subgoal list at the node is gh. Further assume that the search space for g is S_g, and that the search space for h is S_h. By including the additional conjunct h we have replaced every solution node in S_g with the search space S_h. The size of the new search space is proportional to the product of the sizes of the two individual search spaces S_g and S_h. This problem is further exacerbated by additional extraneous conjuncts, as the search space grows exponentially with the number of extraneous conjuncts. Clearly, this combinatoric explosion has a dramatic impact on the efficiency of the problem solving effort.

The most important impact of extraneous conjuncts in the subgoals of a node is that the problem solving process may be incomplete. The additional conjuncts can be inconsistent with the other essential conjuncts in the subgoal due to the constraints of the surrounding components.

Using the reformulated design we can make use of the behavior description of the carry output directly to deduce how to control it to 1 or 0. This requires matching the goal with the output column of the table specifying the behavior of the carry output and using the inputs for the matching row. The search space using the reformulated design for controlling the carry output to 1 and 0 is given in Figure 2.5.

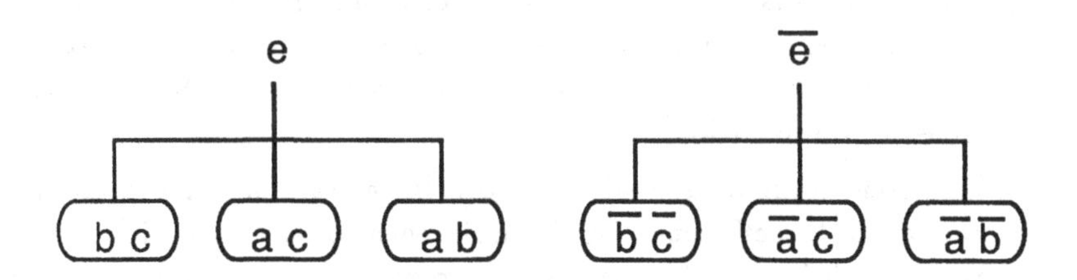

Figure 2.5: Search space for the abstracted full-adder.

Using the reformulated design description we can replace the entire subtree in the original formulation with a single node and its children. All the children in the search space for the reformulated design are solutions, i.e., there are no failure nodes. All the solutions are minimal since there are no duplicate solutions, or extraneous conjuncts in a solution. The behavior of the abstracted design can be minimal since there is no need to share the higher level function specifications of the *sum* and *carry* outputs, although there is a utility in sharing the lower-level implementation of these two functions.

In addition, in abstracting a design, we can capture the *intention* of the designer, which can be lost in the lower-level formulation. These intentions can specify: the encoding of information on a set of connections (defining the unused, or illegal combinations for signals at the lower level), the restricted use of the functionality of the lower-level modules (e.g., in capturing that the *sum* output of the first full-adder of an adder is the exclusive-or of the first two inputs, and that the *carry* output is the *and* of the first two inputs), etc. For these examples, the higher-level behavior would not be defined for any illegal, or unused, argument combinations. In addition, at the abstract level we would define a partial function of the function *generated* by interconnection of the lower-level modules.

In the remainder of this subsection we will evaluate the impact of structural abstraction on the efficiency of the reasoning process and the size of the design. In examining the efficiency of structural abstraction we will compare the relative sizes of the search spaces for a flat low-level design formulation and an abstracted design formulation, including a hierarchical design formulation. We will compare the sizes of the search spaces for the goal of controlling an output of the design to some value (we are only interested in a single solution).

Utility of Structural Abstraction— Increasing Efficiency

Assume that the depth (length of the longest path from an input to an output, ignoring cycles) of the flat low-level design formulation is l, and that the number of different states (cross product of values for state variables) of the design as a whole is s. Further assume that the average number of preconditions in each rule is c (in reducing each goal on the average there will be c new conjunctive subgoals), and that the average branching factor at each node is n (the average number of rules that conclude the same goal). The depth of the search space for the flat formulation is approximately $l \times c \times s$. The size of the search space for the flat formulation is thus:

$$\frac{n^{lcs+1} - 1}{n - 1}$$

Let the number of solutions for the original goal be g. The search space for the abstracted formulation includes the single goal node with g children, each child defining a distinct and minimal solution for the goal. Thus, for the abstracted design we only need to explore a single node. However, the number of nodes to explore for the flat formulation is approximately:

$$\frac{1}{g} \times \frac{n^{lcs+1} - 1}{n - 1}$$

For a hierarchical design with h levels, assume that the average depth per hierarchy level is d, and that the average number of preconditions per rule is c. If we only reason at the lowest abstraction level for this design the depth is d^h. If the number of different states in the design is s, then the depth of the search space, using the lowest level design abstraction, is approximately $d^h \times c \times s$. The number of nodes in the search space for the lowest level of abstraction is thus:

$$\frac{n^{d^h cs+1} - 1}{n - 1}$$

If the number of solutions to the initial goal is g, then the average number of nodes to explore in the search space for the lowest abstraction level is:

$$\frac{1}{g} \times \frac{n^{d^h cs+1} - 1}{n - 1}$$

If we use the search space for the hierarchical design (shifting up one abstraction level at every hierarchy boundary) the depth of the design is $d \times h$, and the depth of the search space is approximately $d \times h \times c \times s$. The size of the search space for the hierarchical formulation is thus:

$$\frac{n^{dhcs+1} - 1}{n - 1}$$

and the average number of nodes to explore before finding a single solution in the search space for the hierarchical formulation is:

$$\frac{1}{g} \times \frac{n^{dhcs+1} - 1}{n - 1}$$

The ratio of the average number of nodes to explore before finding a solution for the flat vs. the hierarchical formulation is:

$$\frac{n^{d^h cs+1} - 1}{n^{dhcs+1} - 1} \approx n^{cs(d^h - dh)}$$

Exploiting the hierarchical design formulation can provide a significant savings since this ratio is exponential in the difference between the depth of flat design and the depth of the hierarchical design. With an explicit representation of the higher level behaviors there is no additional overhead in handling the more complex functions. However, if the higher level functions are axiomatized (described in the next chapter) this benefit is reduced by the additional complexity of reasoning with higher-level functions compared to the complexity of the low-level functions.

Utility of Structural Abstraction— Reducing Design Size

By reformulating designs we can tradeoff between specifying relations directly for all components, or sharing definitions between all similar components. In describing designs, this translates to choosing a space/time tradeoff. This tradeoff is made possible since many of the functions and relations in a design are similar to each other. For example, the functions defining the behavior of all *and* gates in a design are identical to each other, modulo the input/output port names.

We can make explicit, or share, the definitions of both structure and behavior. For example, we can define the behavior of each individual *and* gate in a design by specifying the input/output relationships explicitly for each instance. Alternatively, we can define the behavior of an *and* gate once, and share this definition for all instances. By sharing behavior descriptions we only need to associate the shared description with each instance, instead of duplicating the behavior specification.

Similarly, we can make explicit, or share, the definitions of the substructure for a class of similar or identical components. For example, we can explicitly record all the subparts (gates) of every full-adder in a design. Alternatively, we can define a full-adder prototype and associate this definition with every full-adder in the design without instantiating its substructure. The definition of the prototype specifies the behavior of the full-adder as a whole, the subparts of a full-adder, the interconnections between the subparts, and behavior for the subparts. The prototype definitions can be parameterized to take into account the differences between the different instances, for example, the number of bits at the input of an adder.

Prototype definitions can be nested arbitrarily to define hierarchical designs. For example, a 4 bit adder prototype can be defined in terms of a full-adder prototype, which itself is defined in terms of the *and*, *or*, and *xor* gate prototypes. A hierarchical design need only specify the definition of each distinct module type as a prototype. Such a design would specify a prototype for the device as a whole, a prototype for each distinct subpart of this prototype, and so on down to a primitive set of prototypes. For example, for an adder device with 4 bits inputs, a hierarchical design need only specify the definition of a 4 bit adder prototype, a full-adder prototype, and the three different gate prototypes.

By defining prototypes hierarchically, the number of definitions in a design can be made proportional to the number of distinct module types, instead of the total number of modules in a design. Suppose a hierarchical design has l levels of hierarchy, and on the average each composite module has b different types of submodules. By sharing the definitions of identical prototypes the number of modules in a design is:

$$\frac{b^{l+1}-1}{b-1} \approx b^l$$

If the average number of submodules of a module is c $(c > b)$, then the number of modules for a flat formulation is equal to the number of primitive modules in the hierarchical design, which is equal to c^l. The ratio of the number of modules for the flat and hierarchical design is approximately $(\frac{c}{b})^l$, which is exponential in the number of levels of

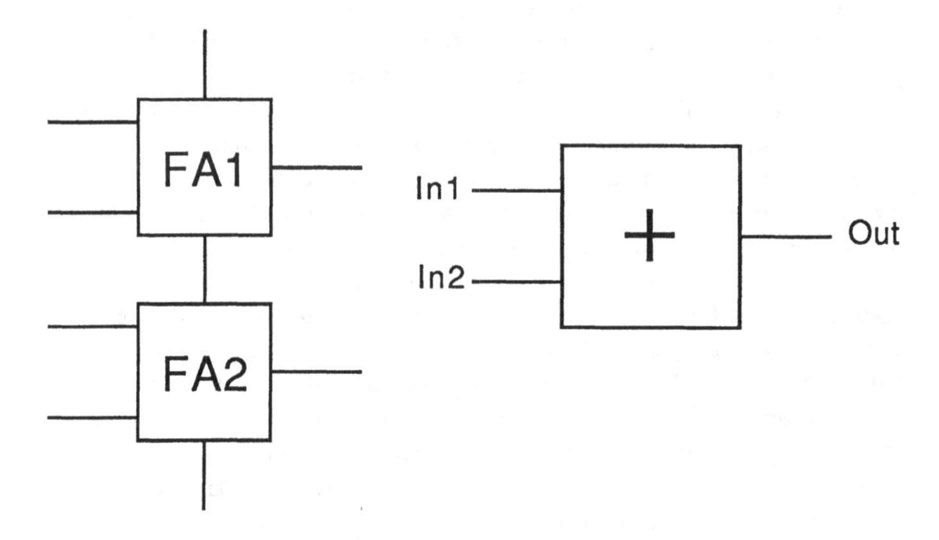

Figure 2.6: An example of spatial abstraction for an adder.

hierarchy, and can be very large if $c \gg b$.

2.2.1.2 Spatial Abstraction

Spatial abstraction corresponds to cases where a single value at the abstract level corresponds to a collection of values at the lower level, each for a different port at the lower level.

An example of spatial abstraction is given in Figure 2.6. The original design is defined in terms of two full-adders connected in a ripple carry fashion, and the new design aggregates these full-adders to define a new adder and its properties. The behavior of the full-adders is defined in terms of boolean values, while the behavior of the adder is defined in terms of integers. A single integer value at an input of the adder maps to two boolean values, one for each of the two subports of each input port. Similarly, an integer value at the output of the adder maps to three boolean values, one for each of the three subports of the output port. The mapping from integers to bits is defined by the standard binary encoding of an integer.

The set of objects in the original design O are the two full-adders, the ports of these full-adders, the 9 connections, and the boolean truth values 0 and 1. The functions in the original design F define the behavior of the *carry* and *sum* outputs of the two full-adders, and the behavior of each connection. Similarly, the relations in the original design R define the type of each component, and define the endpoints of the connections.

The set of objects in the spatially abstracted design include the existing objects in the original design, and the new objects corresponding to the adder, the 3 ports of the adder, the 7 subports of these ports, and the integers from 0 to 6. The new design has four additional functions. The first defines the behavior of the adder as a whole, two others define the mapping between an integer value at the input ports and the two binary values of the subports of each input. Similarly, the last function defines the mapping between an integer value at the output port and the three binary values of its subports. In addition, the new design has 11 additional relations defining the type of the adder, its ports, and the subports of these ports. A formal description of the new design is given below:

$$
\begin{aligned}
O' = O \ \cup \ & \{add, add_{in1}, add_{in2}, add_{out}, 0, \ldots, 6, \\
& add_{in1-0}, add_{in1-1}, add_{in2-0}, add_{in2-1}, \\
& add_{out-0}, add_{out-1}, add_{out-2}, \} \\
F' = F \ \cup \ & \{behav_{add}, behav_{in1}, behav_{in2}, behav_{out}\} \\
R' = R \ \cup \ & \{module(add), \\
& port(add_{in1}, add), port(add_{in2}, add), port(add_{out}, add), \\
& subport(add_{in1-0}, add_{in1}), subport(add_{in1-1}, add_{in1}), \\
& subport(add_{in2-0}, add_{in2}), subport(add_{in2-1}, add_{in2}), \\
& subport(add_{out-0}, add_{out}), subport(add_{out-1}, add_{out}), \\
& subport(add_{out-2}, add_{out}\}
\end{aligned}
$$

The definitions of the functions, modules, and ports are similar to those described in the previous subsection. The relation $subport(add_{in1-0}, add_{in1})$ states that the object add_{in1-0} is a subport of the port add_{in1}, and similarly for the other relations.

The behavior of the full-adders and the adder is given in Table 2.2. These tables directly specify the input/output relations between the ports of the modules. The behavior of full-adders is defined in terms of boolean truth values 0 and 1, and the behavior of the adder is defined in terms of integers in the range 0 to 6. The behavior of the ports is defined similarly by defining the binary equivalent of all integers in the range 0 to 6.

The remainder of this subsection will present an example illustrating the utility of reasoning with the reformulated design of the adder device.

in1	in2	cin	sum	cout
0	0	0	0	0
0	0	1	1	0
0	1	0	1	0
0	1	1	0	1
1	0	0	1	0
1	0	1	0	1
1	1	0	0	1
1	1	1	1	1

Full-adder

in1	in2	out
0	0	0
0	1	1
0	2	2
0	3	3
.	.	.
3	0	3
3	1	4
3	2	5
3	3	6

Adder

Table 2.2: Behavior of the full-adders and the adder.

Suppose the goal is to control the output of the device to 6. Using the reformulated design we can make use of the behavior description of the adder as a whole to find a solution directly. By matching the goal value 6 with the output column of the behavior of the adder we immediately find the solution, i.e., both inputs of the adder must be 3. Thus using the reformulated design there is only a single node in the search space.

Using the original low-level design formulation the goal of controlling the output of the device to 6 is equivalent to the conjunction of following three goals: the carry output of the second full-adder must be 1, the sum output of the second full-adder must be 1, and the sum output of the first full-adder must be 0. Each of these subgoals has a search space associated with it, and we must find the values for the inputs of the device that satisfy each of these subgoals. The size of the search space of the original goal is proportional to the product of the sizes of the search space of the individual subgoals.

In reasoning with the higher level design formulation we are replacing the entire search space at the lower level with a single node and its children. The advantages of reasoning with the reformulated design are similar to those described in the previous subsection for structural abstraction, and the analysis performed there is equally applicable for spatial abstraction.

2.2.1.3 Temporal Abstraction

Temporal abstraction corresponds to the cases where the values at the abstract level are related to the values at the lower level with some delay.

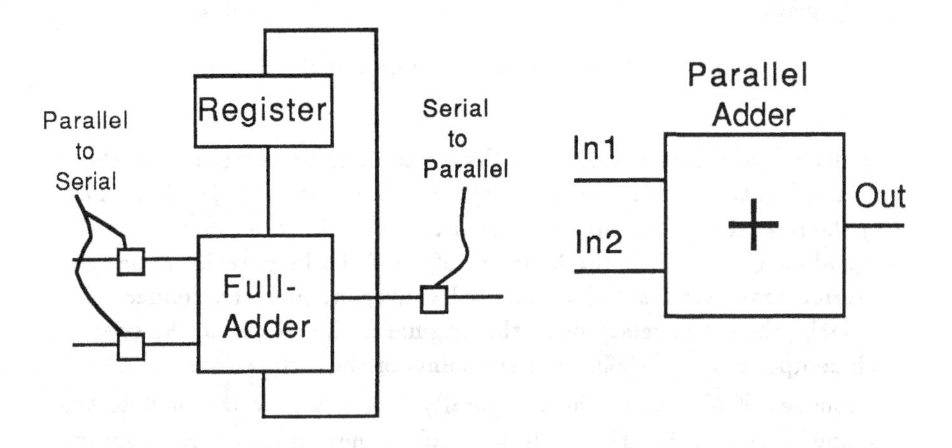

Figure 2.7: An example of temporal abstraction for an adder.

For example, in going from parallel to serial values, a single value for a port at the abstract level at a given time corresponds to a collection of values for its single subport at the lower level, each at a different time. Alternatively, in going from serial to parallel values, a series of values for a port at the abstract level, each at a different time, correspond to a collection of values for its subports, all at the same time.

Such temporal mappings can be common in devices, e.g., the top-level communication can be at the bit-level to minimize the connections between the components, and the internal operations of the modules can be done in parallel for efficiency. Alternatively, the reverse may be true, where the information at the top-level is communicated in parallel via a shared resource, like a bus, and the computations are performed in serial to minimize the hardware.

An example of temporal abstraction is given in Figure 2.7. This figure presents two designs of an adder device which takes integer inputs in the range 0 to 7 and produces the sum of these inputs after a delay of 4 time units. The device is implemented using a serial adder internally. An integer at each input is converted in a stream of 3 bits, and a stream of 4 bits is converted into an integer at the output.

The set of objects in the original design O are the full-adder, the register, the two parallel to serial converters at each input, the serial to

in	out
x@t	x@t+1

Register

in1	in2	out
x@t	y@t	x+y@t+4

Adder

Table 2.3: Behavior of a register and the adder.

parallel converter at the output, the ports of these components, the 8 connections, the boolean truth values 0 and 1, and the set of real numbers (for the integer inputs, and the current time). The set of functions in the original design F define the behavior of the full-adder, register, parallel to serial converters, serial to parallel converter, and the connections. Similarly, the set of relations in the original design R define the type of each component, and define the endpoints of the connections.

The set of objects in the temporally abstracted design include the existing objects in the original design, and the new objects corresponding to the adder, its two input ports, and its output port. The new design has one additional function for the behavior of the adder as a whole. In addition, the new design has 4 additional relations defining the type of the adder and its ports. A formal description of the new design is given below:

$$\begin{aligned} O' &= O \cup \{add, add_{in1}, add_{in2}, add_{out}\} \\ F' &= F \cup \{behav_{add}\} \\ R' &= R \cup \{module(add), port(add_{in1}, add), port(add_{in2}, add), \\ &\qquad port(add_{out}, add)\} \end{aligned}$$

The behavior of the register and the adder is given in Table 2.3. These tables directly specify the input/output relations between the ports of these modules. The entries in the tables specify the value of a port separated by the symbol "@" and the time at which this value is true, e.g., "on@4" denotes a value "on" at time 4. If the register has an input "x" at time "t", then its output will have the same value after a delay of one time unit. The behavior of the other components in the original design are specified without any delay, similar to the examples given in the previous subsections. The behavior of the new adder specifies that its output is the sum of its inputs with a delay of 4 time units.

The remainder of this subsection will present an example illustrating the utility of reasoning with the reformulated design of the adder device. Suppose the goal is to control the output of the device to 12 at time 10.

Using the reformulated design we can make use of the behavior description of the adder as a whole to find a solution directly. By matching the goal value "12@10" with the output column of the behavior of the adder we can find a solution, i.e., we need to find two numbers "x" and "y" such that they add up to 12, and find a number "t" such that "t+4=10". Using the properties of addition (which the system must know) we can find a solution to these goals, e.g., the inputs of the device are 7 and 5 at time 6.

Using the original low-level design formulation the goal of controlling the output of the adder device to 12 at time 10 is equivalent to the conjunction of following four goals: the sum output of the full-adder must be 1 at time 10, 1 at time 9, 0 at time 8, and 0 at time 7. Each of these subgoals has a search space associated with it, and we must find the values and times for the inputs of the device that satisfy these subgoals. The size of the search space of the original goal is proportional to the size of the search space for controlling the sum output of the full-adder raised to the power 4 (one for each subgoal). This can be very large since the lower level design formulation has feedback cycles, and there are potentially an infinite number of solutions to a single subgoal for controlling the sum output of the full-adder (depending on the number of times we cycle through the feedback path).

In reasoning with the higher level design formulation we are replacing the entire search space at the lower level with the task of solving two independent algebraic equations. The advantages of reasoning with the reformulated design are similar to those described in the subsection for structural abstraction, and the analysis performed there is equally applicable to temporal abstraction.

2.2.1.4 Value Abstraction

Value abstraction corresponds to the cases where the set of values at the lower level are partitioned into equivalence classes such that all values in the same equivalence class at the lower level map to a unique value at the abstract level. Value abstraction hides some detail at the lower level, since the set of values at the lower level is larger than the set of values at the abstract level.

An example of value abstraction is given in Figure 2.8. This figure presents two designs of a multiplier device with n bit inputs, and a $2n - 1$ bit output. The most significant bit of the inputs/output is the sign (which is 0 for positive values, and 1 for negative values), and

in1	in2	out
0	0	0
.	.	.
1	2^n	2^n
.	.	.
2^n	2^n	2^{2n}

Multiplier-integers

in1	in2	out
pos	pos	pos
pos	neg	neg
neg	pos	neg
neg	neg	pos

Multiplier- pos/neg

Figure 2.8: An example of value abstraction for a multiplier.

the remaining bits are the magnitude. The original design defines the behavior of the multiplier in terms of integer objects, while the reformulated design specifies the behavior in terms of the objects *pos* and *neg*. The abstraction does not aggregate any structure in the original design, however, it partitions the integers at the lower level into two equivalence classes. All positive integers belong to the first equivalence class and map to the value *pos*, and all negative integers belong to the second equivalence class and map to the value *neg*.

The set of objects in the original design O are the multiplier, its ports, and the integers in the range -2^{2n-2} to 2^{2n-2}. The set of functions in the original design F define the behavior of the multiplier, and the set of relations in the original design R define the type of each component.

The set of objects in the abstracted design include the existing objects in the original design, and the new values *pos* and *neg*. The new design has one additional function for the behavior of the adder in terms of the objects *pos* and *neg*. The new design has no additional relations. A formal description of the new design is given below:

$$O' = O \ \cup \ \{pos, neg\}$$
$$F' = F \ \cup \ \{behav_{mult}\}$$
$$R' = R$$

The function $behav_{mult}$ defines the behavior of the multiplier in terms of the objects *pos* and *neg* as described in the right table in Figure 2.8.

The remainder of this subsection will present an example illustrating the utility of reasoning with the reformulated design of the multiplier device.

Figure 2.9 shows a device made up of two multipliers. The first multiplier has n bit inputs, and the second multiplier has $2n-1$ bit inputs. The most significant bit of an input/output specifies the sign,

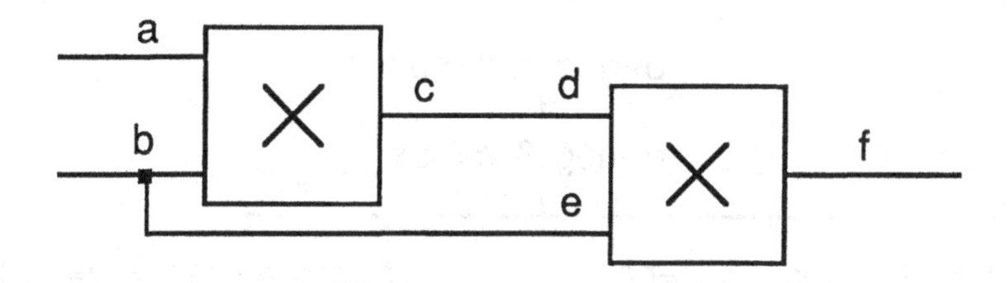

Figure 2.9: A device made up of two multipliers.

and the other bits specify the magnitude.

Suppose we are generating tests for the second multiplier, and the goal is to control the most significant bit of d to 0, and the most significant bit of e to 1. Using the reformulated design we can make use of the behavior description of the multipliers in terms of the objects *pos* and *neg* to achieve this goal. The original goal is equivalent to controlling d to a positive integer, and e to a negative integer. The search space using the reformulated design is given in Figure 2.10. The size of this search space is 5, its depth is 3, and there is one solution node. The average number of nodes to explore before finding a solution is 4.5, and its average branching factor is 4/3.

The search space for the original design formulation in terms of the integer objects is given in Figure 2.11. For this design there are 2^{2n-2} possible positive values for d, and 2^{n-1} possible negative values for e.[4] The search space first branches out to take into account all possible combinations of values for d and e. Similarly, the branches at the end of the search space reflect the different divisors of an integer at the output of the first multiplier. The size of this search space is $n \times 2^{3n-3}$, its depth is 4, and the number of solution nodes is 2^{2n-2}. The average number of nodes to search before finding a solution is approximately $n \times 2^{n-1}$, and the average branching factor is approximately n.

Value abstraction provides a reduction in the average number of nodes to explore which is exponential in the number of bits at each

[4]The inputs to the second multiplier are $2n - 1$ bits wide. However, since e is connected to b which has n bit inputs, the $n-1$ most significant magnitude bits of e are always 0.

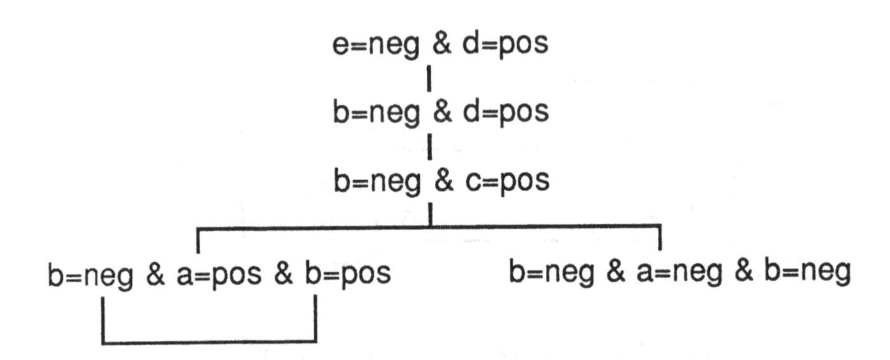

Figure 2.10: Search space using the value abstracted behavior of multipliers.

input of the first multiplier. This reduction is primarily due to the reduction in the branching factor of the nodes in the search space, since the depth of the two search spaces is approximately the same.

In general, the branching factor of the nodes in the search space is reduced when we use an abstract vocabulary for describing the relations between the ports of a module. The terms in the new vocabulary stand for a class of values at the existing abstraction level. For a hierarchical design with branching factor b, the size of the search space is proportional to b^{dhcs}, where d is the average depth of the modules per hierarchy level, h is the number of levels of hierarchy, c is the average number of preconditions in the rules of the design, and s is the total number of states in the design. By reducing the average branching factor to $\frac{b}{k}$ the size of the hierarchical search space is reduced by the factor k^{dhcs}. For large designs, a small reduction in the average branching factor can produce a substantial reduction in the size of the search space.

2.2.1.5 *Function Abstraction*

In the previous subsection we saw how we can extend the vocabulary to define new terms that represent a class of values. Similarly, function abstraction corresponds to the case where we create a new term and associate it with a function by defining its properties. In the reformulated design the behavior of an existing component is described using the new

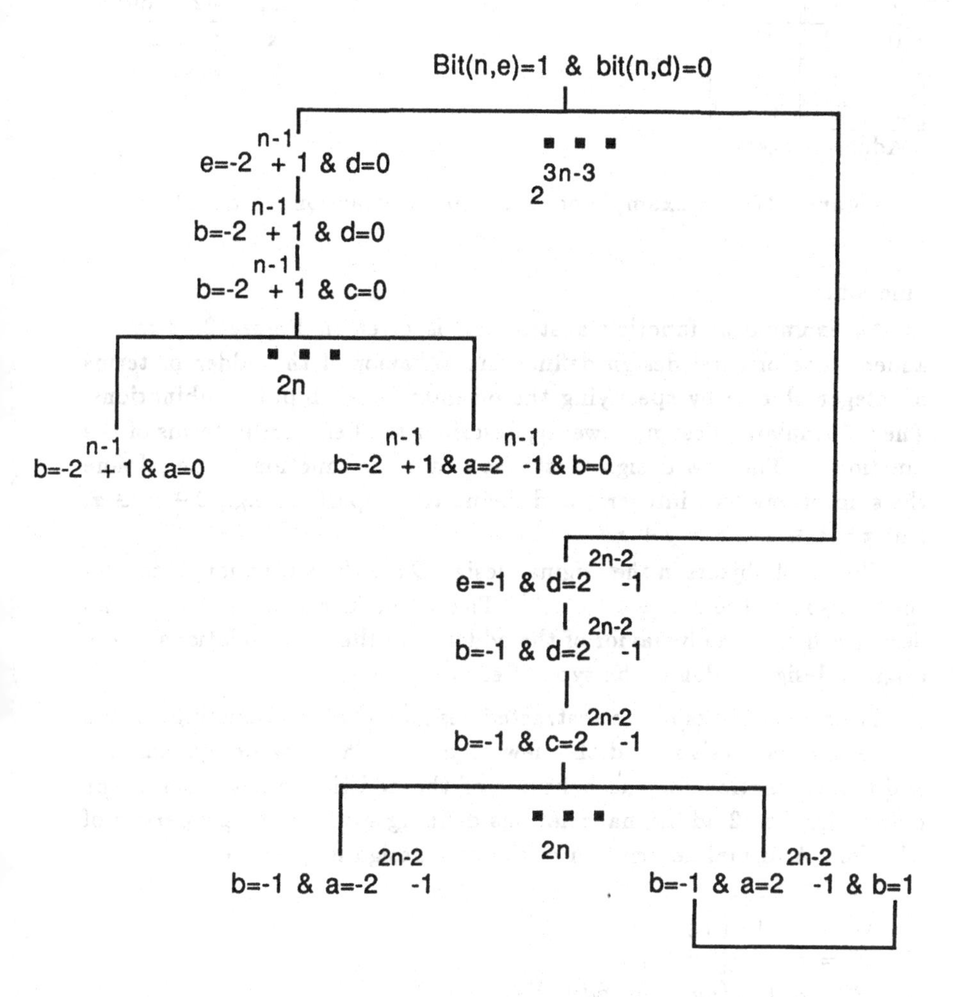

Figure 2.11: Search space using the integer behavior of multipliers.

in1	in2	out
0	0	0
.	.	.
0	2^n	2^n
.	.	.
2^n	2^n	2^{n+1}

Adder-integers

in1	in2	out
x	y	x+y

Adder- +

Figure 2.12: An example of functional abstraction for an adder.

function.

An example of function abstraction is given in Figure 2.12 for an adder. The original design defines the behavior of the adder in terms of integer objects by specifying the outputs for all input combinations. The reformulated design, however, describes the behavior in terms of the function +. The new design must also define the function + (i.e., define the sum of any two integers) and define its properties, e.g., $0 + x \equiv x$, and $x = y \equiv x + z = y + z$.

The set of objects in the original design O are the adder, its ports, and the integers in the range 0 to 2^{n+1}. The set of functions in the original design F define the behavior of the adder, and the set of relations in the original design R define the type of each component.

The set of objects in the abstracted design include the existing objects in the original design, and the new term +. The new design has one additional function for the behavior of the addition function, and the new design has 2 additional relations defining some of the properties of addition. A formal description of the new design is given below:

$$O' = O \cup \{+\}$$
$$F' = F \cup \{behav_+\}$$
$$R' = R \cup \{add\_rel_1, add\_rel_2\}$$

The function $behav_+$ defines the + function by specifying the sum of any two integers. The relation add_rel_1 specifies that $0+x \equiv x$, similarly, add_rel_2 specifies that $x = y \equiv x + z = y + z$.

The remainder of this subsection will present an example illustrating the utility of reasoning with the reformulated design of the adder, and later analyze the utility of functional abstraction in general.

Suppose that the initial goal is to control both inputs of the second

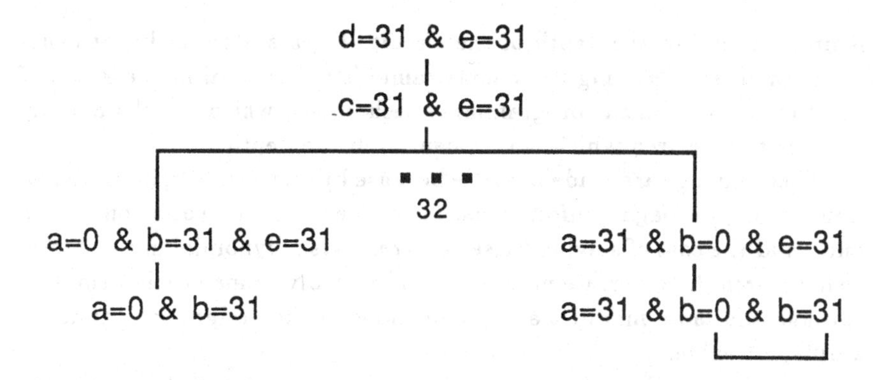

Figure 2.13: Search space for I/O table behavior formulation.

adder to 31, i.e., d=31 ∧ e=31. In the reformulated design the behavior of the first adder is defined as $c = a+b$, and the behavior of the second adder is defined as $f = d+e$. Using this design formulation we can achieve the original goal by propagating and solving symbolic *constraints*, as shown below:

```
d=31 ∧ e=31 ≡ c=31 ∧ e=31
            ≡ a+b=31 ∧ e=31
            ≡ a+b=31 ∧ b=31
            ≡ a+31=31 ∧ b=31
            ≡ a+31=0+31 ∧ b=31
            ≡ a=0 ∧ b=31
```

By propagating symbolic constraints we obtain the solution a=0 ∧ b=31. This solution is obtained without any backtracking, however, it requires simultaneously solving two equations with two unknowns, by taking advantage of the properties of the + function.

In solving the same goal using the original design formulation we have no choice but to employ search, since the behavior of the adders is specified as a table of input/output values. The search space for this design formulation is given in Figure 2.13.

There are 65 nodes in the search space for this design formulation, with only one node having the solution a=0 ∧ b=31. The average number of nodes to explore before finding this solution is thus 65 also.

When employing search to solve a problem, the number of nodes to explore before finding a solution is proportional to the size of the search space which is exponential in the depth of the circuit. However, using constraint propagation the number of equations to solve simultaneously is proportional to the depth of the circuit. For a system of linear equations, the cost of solving these constraints is polynomial in the depth of the circuit using linear-programming techniques, which is substantially cheaper than search which is exponential in the depth.

These savings are made possible because by using constraint propagation we only propagate information through each module and connection once, and the cost of solving these constraints is polynomial in the depth. Using search, however, we may have to repeatedly propagate information through the same modules and connections in the design, each time for a different value.

For a system of non-linear equations, however, constraint propagation can be as expensive as search due to the cost of solving the constraints. In this case it can be more advantageous to employ search if there are good heuristics to guide the selection of promising alternatives at each node in the search space (described in Chapter 4).

2.2.2 Repartitioning Designs

Repartitioning a design involves choosing a different set of objects for a design such that the primitive objects in the new and old design are the same. The reformulated design, thus, defines a new partitioning on top of the primitive set of objects in the old design.

Formally, repartitioning a design involves transforming a design $D = < O, F, R >$ into the design $D' = < O', F', R' >$ such that:

$$\begin{aligned} PrimMod(O) &= PrimMod(O') \wedge \\ PrimFunc(F) &= PrimFunc(F') \wedge \\ PrimRel(R) &= PrimRel(R') \end{aligned}$$

The argument of the function *PrimMod* is a set of modules and its result is the set of modules in its argument that are primitive (i.e., have no substructure), and similarly for the functions *PrimFunc* and *PrimRel* whose result is the set of functions/relations for the primitive modules.

The new set of objects O' can define an alternate structural partitioning of the device, e.g., defining the modules of the design by partitioning a device along its physical boundaries (chips, boards, cabinets, racks, etc.) instead of its functional boundaries (alu, registers, control-store, cache, memory, etc.).

The alternate partitioning can address different concerns in a design. For example, in shifting the partitioning from electrical to physical boundaries we can better address the physical properties of the device, e.g., power consumption, cooling requirements, and packaging for the chips, boards, racks, etc. For example, the system presented in [45] automatically repartitions a functional design hierarchy into a physical design hierarchy. The physical partitioning of a design must take into account the constraints due to the packaging of the functional units in integrated circuits (e.g., four *nand* gates in a SN7400), the constraints of the maximum number of chips per board, the input/output connection limitation per board, the maximum permissible delay per signal, etc. Another example of repartitioning designs is in the system developed by Jin Kim [32] which transforms a functional design partitioning into a physical partitioning for IC cell layout.

2.2.3 Making Design Knowledge Explicit/Implicit

Reformulating a design to make design knowledge explicit/implicit selects a different space time tradeoff. All the facts that could be deduced from the old design can still be deduced in the new design, and vice versa. However, some facts can be deduced more/less efficiently with the new design. The new design either adds, or deletes, redundant functions and relations to/from the old design.

Formally, making design knowledge explicit involves transforming a design $D =< O, F, R >$ into a design $D' =< O', F', R' >$ such that:

$$O = O' \wedge (F \subseteq F' \vee R \subseteq R')$$

The partitioning of the original design is unmodified (the objects in the new design are unchanged), however, we have included additional functions and relations between the objects in the original design. The additional information represents a tradeoff for time at the expense of space, since we are making explicit the relations between the existing objects in the design. These relations are in the deductive closure of the existing design under logical implication.

For example, in achieving a test, the Saturn test generation system caches the solutions to the subgoals of a test. The subgoals for achieving a test include the task of controlling the value of some internal port in the design. The original design does not directly specify how to control the value of this internal port. However, this information can be cached once it is derived by propagating values through the design. Caching involves dynamically reformulating a design during the test generation process by specifying partial functional relationships between the values of the directly controllable inputs and the value of an internal port.

The purpose of caching information is to share computational effort across many tasks. If other tasks require achieving the same subgoals then this information can be looked up directly instead of being recomputed. By caching redundant functions and relations we can improve the efficiency of the reasoning process by a constant factor, at the expense of an increase in the size of the design. The actual increase in efficiency depends on the design and the task. For the previous example this is equal to the average number of tests that require controlling the value of an internal port to the same value. This number is proportional to the depth of the circuit, which can be quite large for complex designs.

Making design knowledge implicit involves forgetting redundant information in the original design. This chooses a different tradeoff in favor of space at the expense of time. Formally, making design knowledge implicit involves transforming a design $D = < O, F, R >$ into a design $D' = < O', F', R' >$ such that:

$$O = O' \wedge (F \supseteq F' \vee R \supseteq R')$$

2.3 Design Correctness

The previous section described the different types of reformulation operations, and showed that by reformulating designs we can increase the efficiency of the reasoning procedure (replacing a subtree in the search space with a node, reducing the branching factor of the nodes in the search space, and using constraint propagation instead of search) and decrease the size of a design. Having the different design formulations related to each other permits shifting from one abstraction level to another, whenever it is computationally advantageous to do so.

However, the computational savings in using the more appropriate formulation cannot come at the expense of correctness. The behavior of each component in the design must be correct, and the behavior of a composite component must be a correct abstraction of the composition of the behaviors of its subparts. However, the two behaviors need not be equivalent, since in defining the behavior of more abstract components we will usually want to ignore the details that are irrelevant at that level. There is a tension between the two competing goals of correctness and ignoring detail. Maintaining correctness must strike a balance between these conflicting goals.

In this section we will define the criteria for a design to be a correct specification of a device. This is based on defining a mapping between the behavior of the device and the behavior specified by the design at each abstraction level. The correctness criteria is extended to ensure that the design specification at the different abstraction levels are consistent with each other.

2.3.1 Correctness for an Abstraction Level

In general, a design is a specification of a device at a collection of abstraction levels. In order for a design to be correct, the specifications at each abstraction level must be a correct specification of the device. The specification of a design at a given abstraction is correct if it correctly specifies both the structure and behavior of the device.

The structural formulation of a design at a given abstraction level is correct if every module, port, connection and state-variable of the design is a part of the device. By *part* we mean a tuple of three-dimensional regions in the device. For example, a port $in1 \in O$ can be a tuple of four physical regions in the device (in this case, an integer value for the port $in1$ is equivalent to four voltages, one for each region). However, the structural formulation of a design need not be *complete*, since we may wish to ignore irrelevant detail (e.g., the power and ground paths, and the plastic casing of a device). A structural formulation is complete if every physical entity in a device is a part of some three-dimensional structure of a module, port, connection or state-variable in the design.

These definitions permit different objects in a design to overlap, i.e., to correspond to overlapping physical regions in the device. In addition, these definitions force a correspondence between the structure of a design and the structure of a device. This decision reflects a choice on our part. It may be possible to reason about a device using a design formulation that has no structural correspondence to it— however, we are not considering these cases.

In addition to the structural correspondence, the behavior of a design at a given abstraction level must be a correct specification of the behavior of the device. The behavior of a design is correct if every function in the design is an *abstraction* of the function *computed* by its corresponding physical structure in the device.

For example, Figure 2.14 shows a picture of an inverter device, the behavior of the inverter at the boolean level, and the actual behavior of the device. At the boolean level we have abstracted the actual voltages

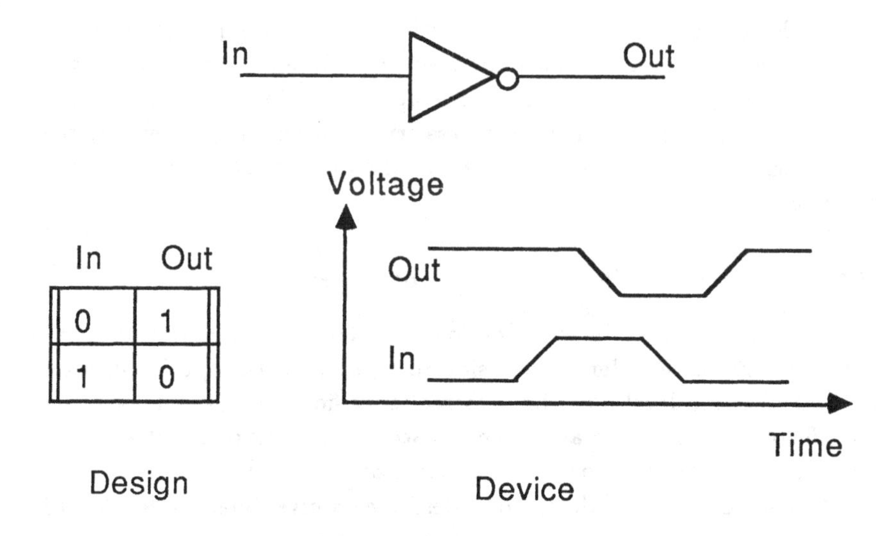

Figure 2.14: Design specification and actual behavior of an inverter.

in the device to the values 1 and 0. All voltages less than or equal to 2.5 volts are interpreted at 0, similarly those greater than 2.5 volts are interpreted as 1. In addition, in the design we have ignored the temporal aspect of the behavior by specifying 0 delay for the inverter.

In order for this design to be a correct specification of the device we must define a homomorphism between the behavior of the device and the behavior specified in the design. This homomorphism is defined using a *commutative diagram*, as shown in Figure 2.15. In this diagram, all paths with the same starting and ending nodes must produce identical results.

The homomorphism is defined by the function h_d which maps a value at the input of the inverter to a value at the input of the design function, and the function h_r which maps a value at the output of the inverter to a value at the output of the design function. There are two important parts of the value of a port— the actual value, and time at which this value is present.

The functions h_d and h_r map a voltage in the range 0 to 2.5 volts to the boolean truth value 0. Similarly, they map a voltage greater than 2.5 volts to the boolean truth value 1. The function h_r maps the time for a voltage at the output of the inverter to the same time for a boolean value at the output of the design. Similarly, the function h_d maps the

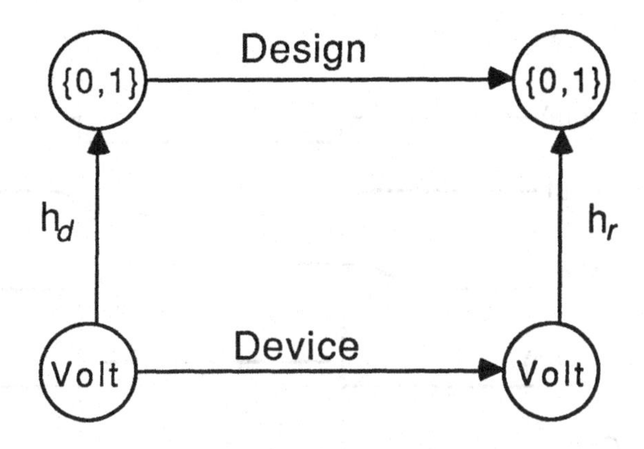

Figure 2.15: Homomorphism between a device and a design.

time for a voltage at the input of the inverter to the same time for a boolean value at the input of the design, except when the voltage at the lower level crosses 2.5 volts. If t is a time at the lower level where the input is crossing 2.5 volts, then all times at the lower level in the range $t \leq time \leq t + delay$ map to the time $t + delay$ at the abstract level (the rise and fall delay of the inverter device is $delay$).

Figure 2.16 illustrates the mapping function h_r for some inputs. This figure shows that the abstraction of the output of the inverter (the top line) is identical to the result of the design for the abstracted inputs (the bottom line).

In general, let the function computed by the device be f', and the function specified in the design be f. The function f' maps arguments in its domain d' to its range r'. Similarly, the function f maps arguments in its domain d to its range r. In general, the homomorphism function h_d maps a collection of values $v1_{t1}, \ldots, vi_{ti}$ from d' (the subscripts denote the time), to a value v_t in d, and similarly for the function h_r. In order for the design to be a correct abstraction of the behavior of the device the following commutative relation must hold true:

$$f(h_d(v1_{t1}), \ldots, h_d(vi_{ti})) = h_r(f'(v1_{t1}, \ldots, vi_{ti})) \qquad (2.1)$$

In general, a design will not *completely* specify the behavior of a device. The behavior of a device includes all functional relationships between arbitrary points in the device. The number of such relation-

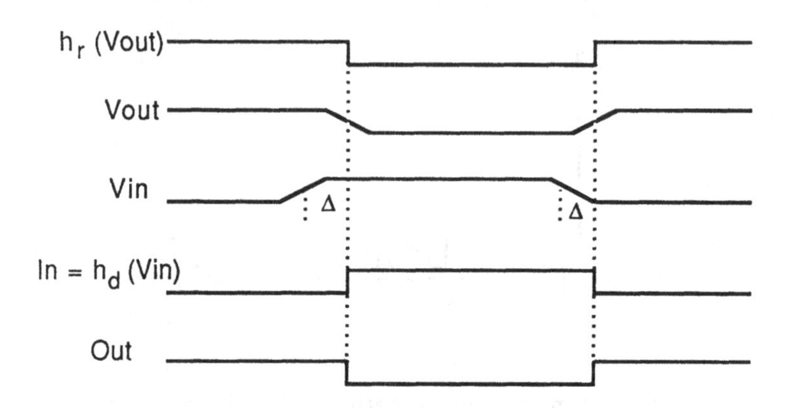

Figure 2.16: Relating the behavior of the inverter and its design.

ships is usually quite large, and it would be impractical to define all of them. Typically, a design formulation will define the more useful behavior relationships, and ignore those that are irrelevant. For example, we can define a partial specification of the behavior for a device by only defining functional relationships for the modules, ports and connections of the design. This formulation omits the relationships between a collection of ports/state-variables from different modules. In addition, the partitioning of the device into a set of modules, ports, connections and state-variables may itself be incomplete.

A device embodies behavior at the level of physics (voltages, electrons, etc). On the other hand a design specifies behavior as a function (possibly a collection of functions). Since a device is a *real* physical entity its behavior is more fine-grained than the approximation of it defined in the design. We can choose the granularity for the definition of behavior in a design by the appropriate selection of the functions h_d and h_r. A design is a more *precise* formulation of a device if the mappings defined by h_d and h_r are more precise. For example, a temporal mapping from a point in time in r' to a point in time in r is more precise than a mapping from a time interval in r' to a time in r (the imprecision is a function of the size of this interval). Similarly, a mapping from a single value in d' to a single value in d is more precise than a mapping from a range of values in d' to a single value in d (again, the imprecision is a function of the size of this interval).

2.3.2 Correctness Between Adjacent Abstraction Levels

In abstracting the behavior of a collection of components at a given abstraction level it is often useful to simplify the descriptions by approximating the exact behavior at the lower level. For a hierarchical design these approximations combine with each other for each level in the hierarchy. We must define the properties of the lowest level behavior that we wish to preserve in order to define if a given design is correct.

We can extend the correctness criteria defined in the previous section between an abstraction level in the design and the device to define the correctness for adjacent abstraction levels in a design. In order for a design specification at a given abstract level to be correct, it must be a correct specification of the structure and behavior of the design at the next lower abstraction level.

The structure of a design at a given abstraction level is correct if every module in the design maps to a collection of interconnected modules at the next lower level. In addition, every component at the lower abstraction level must be a subpart of some (one or more) components at the next higher abstraction level. Similarly, the behavior of a design at a given abstraction level is correct if we can define a mapping between the behavior computed by the subparts of a module and the behavior specified for the module itself, such that the commutative relation specified in Equation 2.1 is true. This mapping is defined by the functions h_d and h_r. In addition, for hierarchical designs, the transitive closure of the homomorphism functions at each abstraction level must satisfy the commutative relation given in Equation 2.1.

The remainder of this subsection will present an example which illustrates the homomorphism between the behavior specifications at adjacent levels in a design.

Figure 2.17 shows two design formulations for the carry function of a full-adder device. The low-level design formulation partitions the device into the interconnection of *and*, or *or*, and *xor* gates, where each gate has a delay of 5 time units. Changes in the inputs that must propagate through the *xor* gate to the carry output have a delay of 15 time units, however, changes that propagate through the lower *and* gate only have a delay of 10 time units.

A structurally abstracted formulation of these gates is given in the

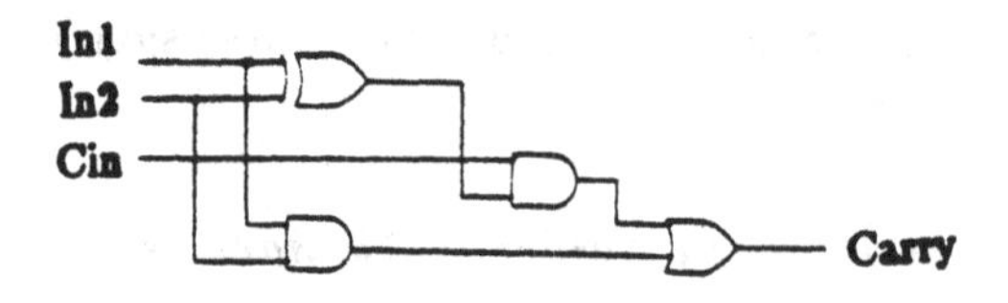

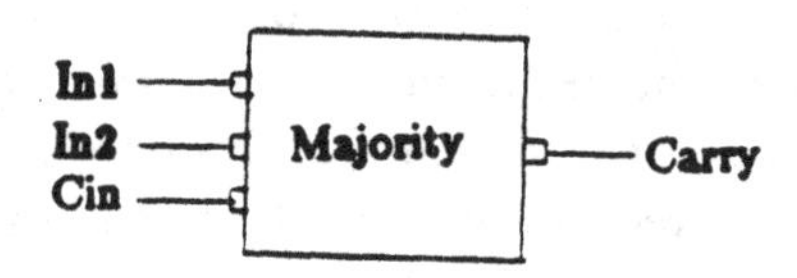

Figure 2.17: Gate level and abstracted design of the carry function.

in1	in2	Cin	Carry
x	0@t	0@t	0@t+10
0@t	x	0@t	0@t+10
0@t	0@t	1@t+5	0@t+15
1@t	1@t	x	1@t+10
0@t	1@t	1@t+5	1@t+15
1@t	0@t	1@t+5	1@t+15

Accurate temporal model

in1	in2	Cin	Carry
x@t	y@t	z@t	Majority(x,y,z)@t+15

Approximate temporal model

Table 2.4: Accurate and approximate specification of the carry function.

right half of the figure as a single three input *majority* function. The subparts of the new module are the existing gates in the original design. To accurately model the behavior for the abstracted module we must enumerate the input combinations, since the delay is a function of the specific values at the input ports. An accurate description of the behavior of the abstracted *majority* function is given in the left half of Table 2.4. This description is given in terms of the function *Majority* which has three arguments, and whose result is the value that is common to two or more of its arguments.

Although this description is accurate, it precludes a symbolic definition (functional abstraction) of the behavior. However, it is possible to provide a symbolic description if we are willing to approximate the exact temporal behavior at the lower level. For example, the right half of Table 2.4 defines an approximate description of the carry function, where the delay is always 15 time units independent of the input values. Consequently, for certain inputs, the abstracted behavior specifies a delay of 5 time units in excess of that specified by the lower level.

Figure 2.18 shows the output of the carry function for the gate level design formulation, and the approximate abstracted design formulation. The inputs for the two design formulations are identical, i.e., they have the same boolean values at the same time.

The behavior of the abstract description of the carry function is correct if we can define the mapping functions h_d and h_r between the gate level and abstract design descriptions such that the commutative relation given in Equation 2.1 is true. The function h_d maps a value v_t at an input at the lower level to the same value v_t at the corresponding port at the abstract level via the identity relation.

In general, in approximating the temporal behavior of a collection of modules at level i, there is a minimum/maximum delay from the inputs of the collection of modules to the output of the collection of modules $\delta_{min,i}/\delta_{max,i}$. An approximation of the temporal behavior for the collection of modules must specify a delay δ_i in this range. The magnitude of the temporal error $\delta_{error,i}$ for any input is bounded by $\delta_{max,i} - \delta_{min,i}$, for abstraction level i. For this example, the minimum and maximum delays at the lower level are 10 and 15, the delay specified at the abstract level is 15, and the maximum temporal error is 5.

For the abstracted design of the carry output the function h_r maps a value v_t at the output of the lower level to the value $v_{abst(t)}$ at the output of the abstract level. That is, a value v at time t at the lower level is mapped to the same value v at time $abst(t)$ at the abstract level. The

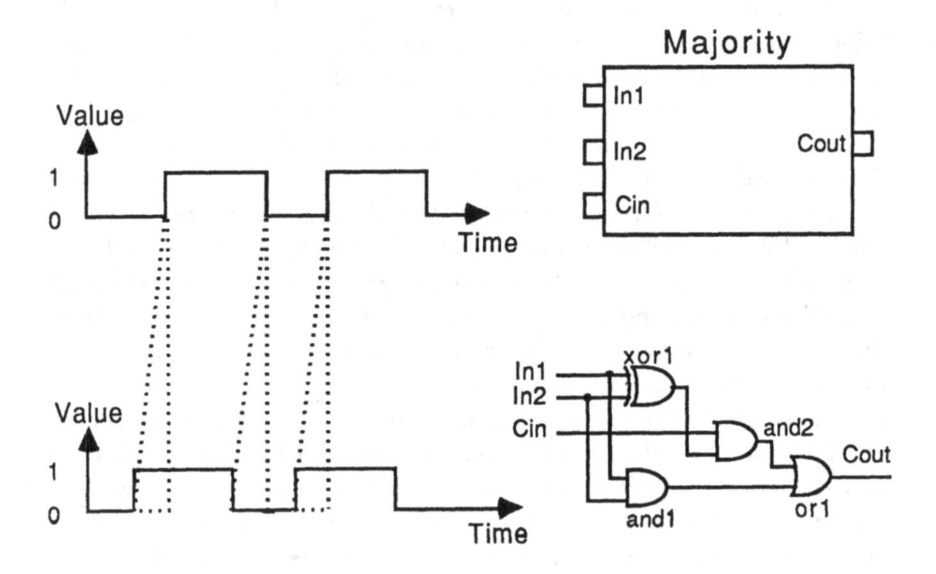

Figure 2.18: Output for the gate level and abstract design formulation of the carry function.

function *abst* maps all times at the lower level to the same time at the abstract level, except that the times at the lower level $t_{transition} < t < t_{transition} + 5$ map to the time point $t_{transition} + 5$ at the abstract level. The expression $t_{transition}$ stands for a time at the lower level at which a transition occurs (from 0 to 1, and vice-versa). The temporal mapping for these exceptions are shown in Figure 2.18 by the dotted triangular regions.

In general, the function h_r must map a time range $t_{transition} < t < t_{transition} + \delta_{level}$ at the lower level to the time point $t_{transition} + \delta level$ at the abstract level. In order for this temporal mapping to be unique these time ranges must not overlap, i.e., the maximum temporal error must be less than the minimum time between transitions ($t_{min,i}$) at the lower level. Therefore, the following constraint must hold true at all abstraction levels for a hierarchical design.

$$\delta_{error,level} < t_{min,level} \tag{2.2}$$

The minimum time between transitions $t_{min,level}$ can be less than the minimum time between transitions for the inputs due to reconvergent fanouts, as is the case, for example, for the device pictured in Figure 2.19 with an inverter connected between the two inputs of an *and* gate. When the input changes from *false* to *true* the output changes from *false* to *true*, and back to *false*. The output maintains the *true* value for a time which is equal to the delay of the inverter, which can be less than the time between input transitions.

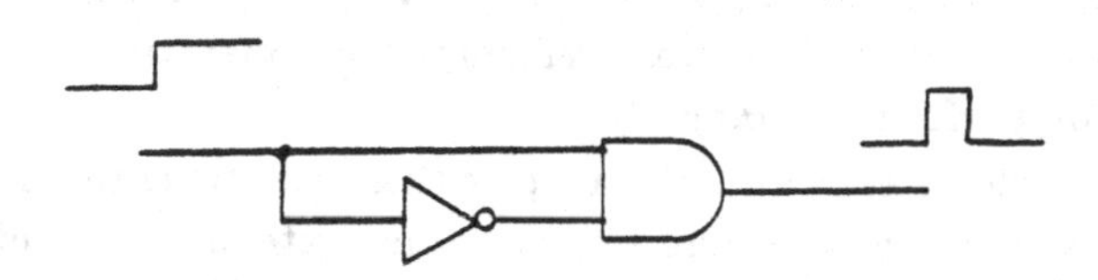

Figure 2.19: Transitions due to reconvergent fanouts.

Sometimes it is useful to add an additional constraint which requires that when a value at an abstract level changes to v at time t, then the values at all the lower levels at the *same* time t must map to this value v (under the transitive closure of the mapping between adjacent levels in the hierarchy). Many reasoning tasks at the higer level are dependent on the time at which the output of a module changes (e.g., simulation and test generation). When using a more abstract description which is an approximation of the exact behavior it is better to restrict the inaccuracies to occur at times other than those at which there are transitions at the higher level.

A consequence of this constraint is that in approximating the delay for a collection of modules we must choose the maximum delay $\delta_{max,i}$, since we must ensure that the output of the high-level behavior does not change before the lower levels. In addition, for the values to agree at the high level transitions we must ensure that the temporal errors at the lower abstraction levels do not add up to greater than the minimum time between transitions. That is, the following constraint must be true:

$$\sum_{level=1}^{i} \left(\delta_{max,level} - \delta_{min,level}\right) < t_{min,i} \qquad (2.3)$$

2.4 Automatically Reformulating Designs

The previous sections have described the different types of reformulation operations, their utility, and the correctness criteria for these reformulations. The goal of performing these reformulations is to increase the efficiency for a given task, and/or reduce the size of the design. In this section we will examine how we can mechanize some of these reformulation operations in order to automatically transform a design closer to its ideal formulation for a given task. In addition, we will show that it is impossible to automate certain reformulation operations, which must, therefore, be performed manually.

It is possible to automate the abstraction operations to reformulate a design. In abstracting a design we can either extend the set of modules, or extend the vocabulary for defining the relations between the ports of the existing modules. In abstracting the set of modules there are two important tasks. We must first select the subparts for each newly created module. These subparts impose a partitioning of the existing modules of the design along some dimension, e.g., their electrically connected boundaries. The second step involves defining the behavior of the newly abstracted modules.

A naive partitioning of an existing design cannot be expected to produce useful abstractions, since small changes in the partition boundaries can lead to a dramatic change in the behavior description for the abstracted modules. For example, consider a collection of gates that implement a four bit adder. By abstracting the entire collection of gates it is possible to define the top-level behavior as out = in1 + in2. However, if a single gate is left out in the partitioning, we are forced to describe the top-level behavior as a large input/output table of boolean values.

One approach to automatically partitioning a design is to use a graph algorithm that partitions a design based on the connectivity of the different modules in the design [17]. This connectivity analysis attempts to partition the design into clusters that are strongly interconnected, and only weakly connected with other clusters. This algorithm can be applied repeatedly to formulate a hierarchical partitioning of a design. It is possible to automatically derive the behavior for a cluster by composing the behavior of the subparts, as is done in the design verification system VERIFY [5]. However, as it was shown in [40], such partitioning algorithms provide poor results since they do not take into consideration other partitioning concerns, e.g., minimizing the complexity of the behavior of the modules, minimizing the number of distinct module types, etc. In other situations where the behavior for the partitions is not important (e.g., in partitioning a design into boards, racks, cabinets, etc.) it is possible to automatically partition a design successfully, as is done in [45].

An alternative is to use a schema-based partitioning of a design, where each schema defines a pattern of components that can be abstracted for a given task. In addition, the schema can specify the behavior for the abstracted modules. We can apply the schema-based abstraction of a design recursively, for example, in abstracting a collection of gates into full-adders, and abstracting the full-adders themselves into an adder. To be useful, it must be possible to parameterize these schemas, e.g., in defining an n bit adder to be made up of a specific interconnection of n full-adders.

We can represent an arbitrary design as a hierarchical graph. The modules of a design correspond to the verteces in the graph, and the relationships between the modules correspond to the edges in the graph (e.g, connections between modules). The substructure of a module is represented by the subgraph of a vertex. We can represent an abstraction schema as a graph similarly. The automatic abstraction task now corresponds to finding subgraphs in the design that match the graph of the schema.

This task is equivalent to the subgraph-isomorphism problem, which is known to be NP-complete [1]. This problem is similar to image understanding in computer vision, where given a set of objects, and the relations between them, the system has to assign an interpretation to the collection of objects (e.g., in identifying a chair from the relative positions of its seat, back, legs, arm-rest, etc.). Numerous algorithms have been proposed in the literature for subgraph isomorphism, e.g., [60] [11] [58] [29]. Fortunately, our task is not as difficult as that for vision, since we are reasoning with reliable data (e.g., no noise, shading, occluding, etc.), and we are not dealing with objects that lack a precise definition. For example, we do not have to label the edges of the schema graph with probabilities to perform probabilistic graph matching, e.g., as would be required for deciding if a cup without a handle is a cup. In addition, we can constrain the subgraph-isomorphism task since each vertex in the design and schema graph is typed (the type of the module), and the edges are labelled by the relations between the modules (e.g., direction of connection). These constraints permit early pruning of inconsistent graph matches when the vertex or edge labellings are not identical.

There can be multiple matches between the schemas and the design, however, some of these abstractions may be mutually inconsistent if they overlap (for a hierarchical design there should be no subparts in common between two modules). We can use an evaluation function to select the most promising abstraction for a collection of mutually inconsistent ones. This is similar to the scheme used in [15] to disambiguate between multiple interpretations of an analog circuit. The scoring function can be based, for example, on removing the abstraction that conflicts with the most other abstractions.

If an abstraction schema does not specify the behavior for the abstracted modules, this behavior must be derived from the behavior of the subparts and their interconnections. Combining the behaviors corresponds to composing the behavior definitions of the low-level modules along their interconnections. For example, in abstracting a collection of gates defining a full-adder we can define the following behavior for the sum and carry outputs (ignoring any delay):

```
sum=xor(xor(in1,in2),cin)
cout=or(and(cin,xor(in1,in2)),and(in1,in2))
```

using the axioms of boolean algebra we can simplify these equations to get:

```
sum=xor(in1,in2,cin)
cout=or(and(in1,in2),and(in1,cin),and(in2,cin))
```

The combination and simplification of the behavior of the subparts is similar to the operations performed in the design verification system *VERIFY* [4].[5] These operations can define a structural hierarchy, i.e., the vocabulary for defining the behavior of the abstracted module is the same as that for its subparts. However, it is not clear how the combined and simplified behavior can be abstracted automatically in the absence of any guidance from the abstraction schemas, e.g., in automating the following operations for abstracting the behavior of a ripple-carry adder made up of 2 full-adders. The behavior for the 3 outputs of the full-adders (the sum outputs of both full-adders, and the carry output of the last) is given below:

$$out_0 = xor(in1_0, in2_0)$$
$$out_1 = xor(in1_1, in2_1, and(in1_0, in2_0))$$
$$out_2 = majority(in1_1, in2_1, and(in1_0, in2_0))$$

Abstracting this behavior would require interpreting the 2 bits $in1_1$ and $in1_0$ as the binary encoding of the integer in1, interpreting the 2 bits $in2_1$ and $in2_0$ as the binary encoding of the integer in2, and interpreting the 3 bits out_2, out_1 and out_0 as the binary encoding of the integer out. With these interpretations, we can prove the following relation between the value of these integers: $out = in1 + in2$.

We can also automatically reformulate designs by making design knowledge explicit. These reformulations can be performed at problem-solving-time as partial solutions to subtasks are computed. For example, we can cache solutions to subtasks, and directly look up these solutions for all tasks that require solving a subtask that has already been solved.

Certain reformulations are impossible to automate, for example, refining a design. In refining a design each newly created module must correspond to some physical region in the device. Since the device is inaccessible to the reasoning machinery it is impossible to automatically verify the structural consistency between the design and the device. Usually there are many different ways of implementing an abstract module, e.g., we can implement an adder as a ripple-carry adder, a carry-look-ahead adder, a table-lookup adder, or a bit serial adder. Without examining the device we having no justification for selecting any particular refinement.

---

[5] An additional operation in this system is to check the equivalence between the top-level behavior and the combined and simplified behavior of the subparts.

However, it is possible to automatically refine a design if the device does not exist, and the design is being refined to a level where it would be possible to fabricate the device directly (the structural correspondence must hold once the device is created). This is equivalent to the automatic design synthesis problem which has been studied by others in the digital domain [57] [41].

It may be impossible to automatically shift the partitioning of a design by examining it along a different dimension, e.g., in partitioning a hierarchical design along its physical boundaries, when it was originally partitioned along its functional boundaries. The original hierarchical design can be flattened, but we have no clue as to the physical proximity of the primitive modules unless this information is already present. Since the device is inaccessible to the reformulation machinery this information cannot be derived. Again, if the device does not exist it is possible to repartition the design along a new dimension as long as the new partitioning is a part of the specification for fabricating the device [32] [45]. Choosing a different partitioning along an existing dimension (physical, logical, etc.) is similar to abstracting a design after flattening it, and this can be automated.

2.5 Manually Reformulating Designs

The previous section described how certain reformulation operations can be automated in order to transform a design closer to its ideal formulation for a given task. In this section we will show that in certain situations it is more advantageous to reformulate a design manually, and that a useful design formulation to consider initially for all tasks is the one created in the design process. This formulation is a compromise between the ideal design formulation and a flat low-level design formulation that corresponds to the detailed structure of the device.

In general, automatically reformulating designs is very difficult. Even in situations where it is possible to automatically reformulate designs, manual reformulation usually produces *better* (this depends on the task) results [40]. In other situations it may be appropriate to automate only some of the reformulation operations. For example, in the system proposed by Richard Korf [33] problem solving is viewed as a heuristic search in the space of problem representations. The user navigates through this search space by selecting the transformations to apply next, and the system generates the reformulated descriptions.

In the digital domain, automatically abstracting a design corresponds to reverse engineering, i.e., deducing an organization that may have been used in designing the device, but which was not included in the final design. Instead of automatically abstracting a flat design description, it is more appropriate to capture the organization of the design as it is refined from the more abstract formulations by the designer. Unfortunately, this information is usually recorded on the backs of envelopes, or left in the mind of the designer, since most traditional design tools force the specification of a design at a particular abstraction level, e.g., transistor level, gate level, or register transfer level. However, this situation is rectified in modern computer-aided design environments, e.g., *Helios* [21] and *Palladio* [9], where the user can choose an arbitrary abstraction level for defining a design, and different parts of the design can be refined to different abstraction levels.

The best formulation for a design is both domain and task dependent. For example, to improve the efficiency of test generation, the best organization should define the value of every internal port as a function of the directly controllable inputs, and define the value of a directly observable port as a function of this port and the directly controllable inputs. However, the hierarchical formulation normally created in the incremental design refinement process is a useful compromise between a flat formulation and the ideal formulation described above. The advantage of the hierarchical formulation is that it is available directly as a result of the design process. In addition, the hierarchical formulation defines fewer modules, and consequently requires less space to represent. For example, suppose that a hierarchical design has l levels, and that on the average the number of modules per level is n. The total number of modules for this formulation of the design is:

$$\frac{n^{l+1} - 1}{n - 1}$$

Similarly, the number of primitive modules at the lowest abstraction level is n^l. Let the number of ports, on the average, for these primitive modules be p. Then the number of modules needed for the ideal formulation is:

$$2 \times p \times n^l$$

In the ideal formulation we have defined two modules per primitive port— one defining the value of this port as a function of the directly controllable inputs, and the other defining the value of a directly observable output as a function of the value at this port and the directly controllable inputs. The ratio between the number of modules for the ideal and hierarchical formulations is $2 \times p$ for large n, and p for $n = 2$.

Each of the modules for the ideal formulation can only be used for subtasks related to controlling, or making observable, that specific port, i.e., the definitions are not shared between different subtasks. On the other hand, the behavior of the modules defined in the hierarchical formulation is shared across multiple tasks at the same, or lower, abstraction levels. For example, in generating tests for the subparts of a full-adder the behavior of the first exclusive-or gate is used in achieving the tests for the second exclusive-or gate, the second *and* gate, and the *or* gate.

The hierarchical formulation produced in the design process is better than a flat low level design formulation that corresponds to the detailed structure of the device. By organizing a design hierarchically we can improve the efficiency and reduce the size of the design, as illustrated in the previous section on abstracting designs.

The reformulation operations that cannot be automated must be performed manually by the user (designer). For example, for test generation, the user should refine the design to the level at which the fault models are relevant. Performing these refinements improves the quality of the tests generated, in addition to improving the efficiency (this will be described further in Chapter 4). The user should also provide any additional information that might help in reasoning about the device, e.g., specifying the physical proximity of the wires allows testing for bridge faults more efficiently, since we only need to test for faults between adjacent wires, instead of between all pairs of wires.

A useful reformulation strategy is to capture as much information from the designer as possible during the design of a device, and to integrate other information by automatically reformulating the design as necessary.

Chapter 3

General Representation and Reasoning

The previous chapter defined the different types of reformulation operations, and showed that by reformulating designs we can increase the efficiency of the reasoning process and reduce the size of the design. A design is a specification of a device at the knowledge level independent of any symbols. In order to reason about a device we must encode its design in a representation language, and use an inference procedure to reason with this description.

In this chapter we will describe general methods based on logic for representing and reasoning about designs. We will first describe the requirements for a design description language. We will next present the syntax and semantics of a language based on predicate calculus, and show how we can describe the structure and behavior of a design using this language. In addition, we will present a complete inference procedure for reasoning about logical descriptions of designs, and illustrate how we can use this inference procedure to simulate, diagnose, and test a device. In describing the reasoning procedure for these tasks we will ignore the control issues (a detailed discussion of this for test generation can be found in the next chapter). Finally, we will describe the utility of using general methods for representing and reasoning about designs.

3.1 Requirements for a Design Description Language

When we use the term *language* we mean a vocabulary for describing designs at the symbol level, and not the actual data structures that are used to encode sentences of the language in a machine. In addition, we are not considering the vocabulary presented to a user to describe designs, but rather the vocabulary *used* by the reasoning mechanism of the language.

In representing a design, the language used must have a well defined syntax and semantics associated with it. The syntax of a language defines the set of sentences (sequence of symbols) that are in the language, and the semantics defines the meaning of any sentence in the language. In describing designs, the semantics of the language must map a well-formed sentence in the language to either an object in the device, a function between the objects of the device, or to a relation between the objects in the device. In addition, the language must define the inference procedures for reasoning with these descriptions. The totality of the knowledge represented in a design description is the union of the facts that are defined explicitly in the description, and those that can be derived from these explicit descriptions using the reasoning mechanisms provided by the language.

A representation language must be general enough to describe an arbitrary design $D = < O, F, R >$. It should be able to represent arbitrary objects, and arbitrary functions and relations between these objects. In order to be a representation language it must have assertional force, i.e. be able to *assert* relations between objects.

In addition, a representation language must be task independent, i.e., it should not rely on any particular application for which the knowledge will be used (e.g., simulation, test generation, diagnosis, etc.). For example, a procedural representation of the functions in a design is adequate for simulation, but not diagnosis. For simulation we are interested in propagating information from the inputs of modules to their outputs. Most devices are functional, therefore, this corresponds to computing the result of a function for given arguments. For diagnosis, however, we are also interested in finding the inputs of a module that could produce a given output. This requires inverting functions, which is extremely difficult if functions are represented as procedures in traditional programming languages[1].

We can characterize the utility of a language satisfying these requirements along a collection of dimensions, e.g., its perspecuity, its efficiency, and its ability to encode incomplete designs evolving over time. By perspecuity we mean the ease with which a design can be encoded in the

[1]At the very least functions would have to be first-class objects, as they are in Lisp and assembly languages. However, inverting functions is made difficult by side-effects, and non-local variable references in the body of a function.

language, e.g., the length of the description, and the ease with which it can be described and understood (the naturalness of the descriptions). The efficiency of a language is the measure of space and time required in representing and reasoning about a design. This is not an absolute measure on some implementation of the language, but rather a measure in terms of the length of the descriptions, the number, and complexity of the inferences required in reasoning about a design (which is related to the set of primitive functions defined in the language).

3.2 Syntax and Semantics for Predicate Calculus

The important syntactic constructs in predicate calculus are constants, variables, terms and propositions. There are three types of constants: object constants, function constants and relation constants. Variables are either quantified universally or existentially. The terms in the language correspond to object constants, variables, or the application of a function constant to a tuple of other terms. There are three types of propositions: atomic propositions, logical propositions and quantified propositions. The sentences in the language correspond to propositions. A formal description of the syntax of a language based on predicate calculus is given in Table 3.1.

The syntax defines the legal sequence of symbols in the language, without assigning any meaning (truth value) to them. On the other hand, the semantics of the language assigns a truth value (*true* or *false*) to every legal proposition in the language. In defining the semantics of the language we will define the objects, functions, and relations that the constants and variables of an expression in the language denote. Having done this we can assign a truth value to structures, functionals, and propositions in the language.

The semantics of the language is defined by the function $\models$ which assigns meaning to every legal expression in the language. This function is defined using the set U which is the universe of discourse, i.e., the set of objects we are concerned with (e.g., the parts of the device and the values used to describe its behavior). The semantic function $\models$ defines a proposition p to be *satisfied* (be *true*) for a given interpretation I and assignment A. This is written as $\models_I p[A]$. The function $\models$ is defined in terms of the function Φ which maps the terms to elements and tuples from the set U, and function and relation constants to a set of n-tuples from the set U (for a given interpretation). The interpretation function I defines a mapping from constants to elements of the set U, or to a

```
<const>            ::=  <obj-const>|<func-const>|<rel-const>
<obj-const>        ::=  <Char-string>
<func-const>       ::=  <Char-string>
<rel-const>        ::=  <Char-string>
<Char-string>      ::=  <Char>|<Char><string>
<string>           ::=  <char>|<char><string>|<Char>|<Char><string>
<char>             ::=  a|b|c|...
<Char>             ::=  A|B|C|...
<variable>         ::=  <char-string>
<char-string>      ::=  <char>|<char><string>
<structure>        ::=  [<term-list>]
<term>             ::=  <obj-const>|<structure>|<variable>|
                        <func-const>(<term-list>)
<term-list>        ::=  e|<term-listp>
<term-listp>       ::=  <term>|<term>␣<term-listp>
<atomic-prop>      ::=  (<rel-const>)|(<rel-const>␣<term-listp>)|
                        (=␣<term>␣<term>)
<logical-prop>     ::=  ¬<proposition>|<proposition>∨<proposition>|
                        <proposition>∧<proposition>
<quantified-prop>  ::=  ∀<variable><proposition>|∃<variable><proposition>
<proposition>      ::=  <atomic-prop>|<logical-prop>|<quantified-prop>
```

Table 3.1: Syntax for Predicate Calculus.

$$\begin{array}{lll} A: & \text{<variable>} & \rightarrow U \\ I: & \text{<obj-const>} & \rightarrow U \\ & \text{<rel-const>} & \rightarrow \subseteq U^n \\ & \text{<func-const>} & \rightarrow \subseteq U^{n+1} \end{array}$$

Table 3.2: Mappings for the Interpretation and Assignment functions.

$$\begin{array}{l} \Phi_{I,A}(\text{<variable>}) = A(\text{<variable>}) \\ \Phi_{I,A}(\text{<const>}) = I(\text{<const>}) \\ \Phi_{I,A}([\text{<term1>} \ldots \sqcup \text{<termi>}]) = < \Phi_{I,A}(\text{<term1>}) \ldots \Phi_{I,A}(\text{<termi>}) > \\ \Phi_{I,A}(\text{<func-const>}(\text{<term1>} \ldots \sqcup \text{<termi>})) = \\ \quad \alpha_j \iff < \Phi_{I,A}(\text{<term1>}) \ldots \Phi_{I,A}(\text{<termi>})\ \alpha_j > \ \in \ \Phi_{I,A}(\text{<func-const>}) \end{array}$$

Table 3.3: Definition of the semantic function Φ for terms.

set of n-tuples from the set U. The assignment function A defines a mapping from variables to elements of the set U. The functions defining the semantics for variables and constants are given in Table 3.2.

The interpretation function I maps a relation constant of n arguments to a set of n-tuples from the set U. Similarly, it maps a function constant of n arguments to a set of $n+1$-tuples from the set U. The first n elements of a tuple correspond to the arguments of the function, and the last element corresponds to the result (there cannot be tw, or more tuples in this set with the same value for the first n elements). The definition of the function Φ, which defines the semantics for terms, is given in Table 3.3.

Finally, the definition of the *satisfies* predicate for propositions ($\models$), for a given interpretation and assignment is given in Table 3.4.

An interpretation I is a *model* of a proposition p if I satisfies p for all assignments (written as $\models_I p$). Similarly, an interpretation I is a model of a set of propositions T if I satisfies every $p \in T$ for all assignments. A proposition p is *valid* if every interpretation is a model of p, i.e., $\models_I p$ for all I. A proposition s *logically implies* a proposition p (written as $s \models p$) if for every interpretation I, $\models_I s \Rightarrow \models_I p$ (this can be extended to a set of propositions logically implying another). A *theory* is a set of propositions closed under logical implication, i.e., for a set of propositions T, $T \models_I p \Rightarrow p \in T$.

$$\models_I(=\sqcup\text{<term1>}\sqcup\text{<term2>})[A] \iff$$
$$\Phi_{I,A}(\text{<term1>}) = \Phi_{I,A}(\text{<term2>})$$

$$\models_I(\text{<rel-const>}\sqcup\text{<term1>} \ldots\sqcup \text{<termi>})[A] \iff$$
$$< \Phi_{I,A}(\text{<term1>}) \ldots \Phi_{I,A}(\text{<termi>}) > \in \Phi_{I,A}(\text{<rel-const>})$$

$$\models_I \neg\text{<proposition>}[A] \iff \not\models_I\text{<proposition>}[A]$$

$$\models_I\text{<proposition1>}\wedge\text{<proposition2>}[A] \iff$$
$$\models_I\text{<proposition1>}[A] \text{ and } \models_I\text{<proposition2>}[A]$$

$$\models_I\text{<proposition1>}\vee\text{<proposition2>}[A] \iff$$
$$\models_I\text{<proposition1>}[A] \text{ or } \models_I\text{<proposition2>}[A]$$

$$\models_I \forall\text{<variable><proposition>}[A] \iff$$
$$\models_I\text{<proposition>}[A'] \text{ for all } A' \text{ and } \alpha \in U$$
where for any α, A' is defined by:
$$A'(\text{<variable>}) = \alpha$$
$$A'(\text{<variable'>}) = A(\text{<variable'>}) \text{ for all <variable'> } \neq \text{ <variable>}$$

$$\models_I \exists\text{<variable><proposition>}[A] \iff$$
$$\models_I\text{<proposition>}[A'] \text{ for some } A' \text{ and } \alpha \in U$$
where for some α, A' is defined by:
$$A'(\text{<variable>}) = \alpha$$
$$A'(\text{<variable'>}) = A(\text{<variable'>}) \text{ for all <variable'> } \neq \text{ <variable>}$$

Table 3.4: Semantics for predicate calculus.

These definitions provide a semantic basis for deciding if a set of propositions P is logically implied by another set S ($S \models P$).

The truth of a proposition in predicate calculus depends on the interpretation of the symbols in the language (I). A computer cannot reason about the truth of a proposition, it must be done outside the computer since the interpretation function I maps symbols (in a computer) to objects in the *real* world. For example, (module and1) is an atomic proposition where the interpretation of the symbol module is the set of 1-tuple objects that are modules. Similarly, the interpretation of the object constant and1 is an object from the set U. This proposition asserts that the object and1 is a module. To check the truth of this proposition we must check if the object corresponding to the symbol and1 is a module in the *real* world.

In describing a design in predicate calculus we will define an interpretation I which is a model for a set of propositions T for the design. In order to reason effectively with the design description, the set of proposition T will usually make explicit the definition of the interpretation for relation constants and function constants. Every relation constant corresponds to a set of n-tuples, where the components of the tuples are elements of the set U. We can make explicit the interpretation of a relation constant by including an atomic proposition for each n-tuple of the relation in T. For example, we can include (module and1) for the tuple $< and1 >$, which is a member of the set of tuples in the interpretation of the relation constant module. This is repeated for all tuples of a relation constant, and all relation constants. Similarly, each function constant corresponds to a set of $n+1$-tuples, where the components of the tuples are elements of the set U. We can make explicit the definition of a function constant by including an atomic proposition for each $n+1$-tuple of the function. For example, we can include (= 0 and(0 1)) for the tuple $< 0\ 1\ 0 >$, which is a member of the set of tuples in the interpretation of the function constant and. Again, this is repeated for all tuples of a function constant, and all function constants.

3.3 Describing Designs

In this section we will represent the structure and behavior of a design in the language Corona [51], which defines a vocabulary for describing

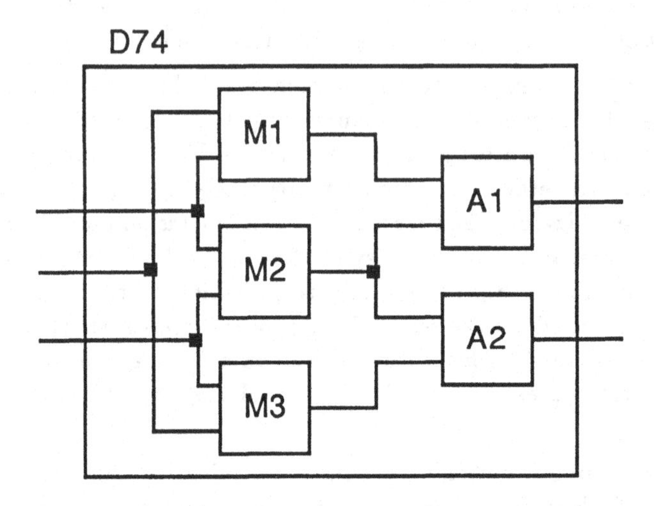

Figure 3.1: A picture of the structure of a design for the device D74.

designs in predicate calculus.[2] The vocabulary defines a naming convention for various function and relation constants, without defining their interpretation. For example, the vocabulary can specify that the symbol "module" is to be used for defining the physical components in the set U, instead of the symbol "#$34!". In addition, CORONA defines a set of primitive functions which can be used in defining other functions (e.g, $+$, $-$, $\times$, and $\div$).

Describing a design $D = < O, F, R >$ involves defining the universe of discourse U (which is the same as the set O), choosing a set of constant symbols for the object, function and relation constants and defining the interpretation function I for these, and making explicit the interpretation of the relation and function constants via a set of propositions.

As an example, we will present the description of a design for the device D74, which was described earlier in section 2.1.2. A picture of the structure of a design for this device is given in Figure 3.1. In this design we have partitioned the device D74 into the interconnection of 3 multipliers M1, M2, and M3, and 2 adders A1 and A2. The device as a whole has three inputs and two outputs. In this design we have chosen to ignore

[2]The syntax of the descriptions corresponds to Table 3.1, which is slightly different from that of Corona.

A0:	(Polybox D74 D74-in1 D74-in2 D74-in3 D74-out1 D74-out2)
A1:	(Mult M1 M1-in1 M1-in2 M1-out)
A2:	(Mult M2 M2-in1 M2-in2 M2-out)
A3:	(Mult M3 M3-in1 M3-in2 M3-out)
A4:	(Adder A1 A1-in1 A1-in2 A1-out)
A5:	(Adder A2 A2-in1 A2-in2 A2-out)
A6:	(Conn D74-in1 M1-in2)
A7:	(Conn D74-in1 M2-in1)
A8:	(Conn D74-in2 M1-in1)
A9:	(Conn D74-in2 M3-in2)
A10:	(Conn D74-in3 M2-in2)
A11:	(Conn D74-in3 M3-in1)
A12:	(Conn M1-out A1-in1)
A13:	(Conn M2-out A1-in2)
A14:	(Conn M2-out A2-in1)
A15:	(Conn M3-out A2-in2)
A16:	(Conn A1-out D74-out1)
A17:	(Conn A2-out D74-out2)

Table 3.5: Structural description for D74.

the power (+5, and 0 volts) pins, and the network of wires connected to these in the integrated circuit. In addition, we have simplified the behavior of each component by assuming 0 delay. We are also ignoring the physical aspects of the device in this partitioning, e.g., the relative physical positions of the parts.

The set of objects O for this design are: the 3 multipliers and 2 adders, the ports of these modules, the 12 connections, and the set of integers (for describing the behavior). There are no state-variables for this design. The set of functions F defines the behavior of each module and connection in the design, and the set of relations R defines the type of each module, and the endpoints of each connection.

In describing this design in predicate calculus, the universe of discourse U is the same as the set of objects O. Each object constant will be denoted by a unique symbol (whose interpretation is the object). For example, the numeral 1 for the number one, and the symbol D74-in1 for the first port of the module D74. Each function in the design will be denoted by a function constant whose interpretation is a set of $n+1$-tuple of objects (the first n elements of the tuple are the arguments, and the last element is the result of the function for these arguments). Each function is associated with one or more components, e.g, the function constant $\times$ is associated with the 3 multipliers in the design.

B1: ¬(Mult x i1 i2 o) ∨ ¬(= a Val(i1)) ∨ ¬(= b Val(i2)) ∨ ¬(= c (× a b)) ∨ (= c Val(o))

B2: ¬(Adder x i1 i2 o) ∨ ¬(= a Val(i1)) ∨ ¬(= b Val(i2)) ∨ ¬(= c (+ a b)) ∨ (= c Val(o))

B3: ¬(Conn x y) ∨ ¬(= a Val(x)) ∨ (= a Val(y))

Table 3.6: Behavior description for D74.

The description of the structure of this design in predicate calculus is given in table 3.5. The fact A1 asserts that the module M1 is a multiplier with ports M1-in1, M1-in2, and M1-out, and similarly for the other modules. The remaining facts define the endpoints of the connections. For example, the fact A17 asserts that port A2-out is connected to port D74-out2.

The behavior of the components is defined in conjunctive normal form (CNF), i.e., by a conjunction of clauses, each of which is a disjunction of literals. A literal is an atomic proposition (no logical connectives), or the negation of an atomic proposition. It is possible to transform arbitrary propositions into CNF [43]. For example, a rule of the form a ∧ b ⇒ c is equivalent to clause ¬a ∨ ¬b ∨ c in CNF, if a, b, and c are atomic propositions.

The description of the behavior of this design in CNF is given in table 3.6. The terms starting with lower case letters in these descriptions stand for universally quantified variables, e.g., x.[3] The fact B1 defines the behavior for all multipliers in terms of the function ×, which is predefined in Corona. This fact asserts that if x is a multiplier with inputs i1 and i2 and output o, and the value of i1 is a, and the value of i2 is b, and the product of a and b is c, then the value of o is c. Similarly, the second fact defines the behavior for all adders, and the last fact defines the behavior for all connections, which are unidirectional with 0 delay.

3.4 Automated Deduction

The previous section showed how to represent arbitrary designs in predicate calculus. In order to be useful, it is important to be able to reason with these design descriptions. In this section we will present a collec-

[3] A fact of the form (= c val(A1-in1)) is equivalent to ∀c(= c val(A1-in1)). We have not explicitly recorded the quantification of variables to increase the readability of these descriptions.

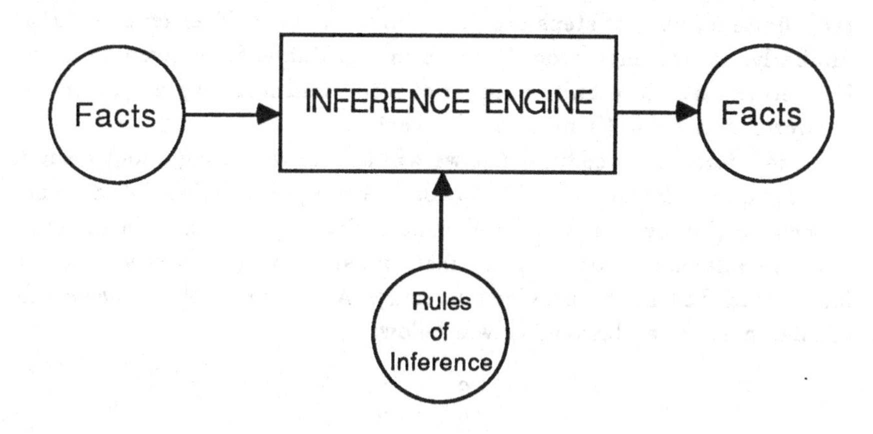

Figure 3.2: The automated deduction process.

tion of inference procedures which can be used to reason about designs. We will illustrate the use of these inference procedures for simulating, diagnosing, and testing designs. Finally, we will discuss the importance of controlling the automated deduction process in order to increase efficiency.

Given a collection of facts, we would like to reason about them to draw further conclusions. For example, given a fact defining the behavior of a connection, and another fact asserting that the starting port of the connection has the value 1, then we would like to be able to conclude that the ending port of the connection also has the value 1.

An inference procedure is an automated deduction method for reasoning about a collection of facts in order to draw further conclusions. A picture of the automated deduction process is given in Figure 3.2. A general inference engine takes as inputs a collection of facts and the description of an inference procedure, and deduces a separate set of facts. In automating the deduction process we would like to use an inference procedure that is sound and complete.

The statement $S \vdash P$ asserts that the set of propositions P is proved from the set of propositions S by a given inference procedure. An inference procedure is *sound* (preserves correctness) if $S \vdash P \Rightarrow S \models P$. Similarly, an inference procedure is *complete* (can actually find all the logical implications by running this method) if $S \models P \Rightarrow S \vdash P$. In addition, an inference procedure is decidable, for a given problem, if after a finite number of steps the procedure indicates either *true* or *false*. Similarly, an inference procedure is semi-decidable, for a given problem, if it returns *true* in a finite number of steps, and if it may not return an answer if it is *false* (it may run forever).

The inference procedure that we will use for reasoning about designs is based on resolution [44]. Resolution is a complete and sound inference procedure (however, it is semi-decidable for implication). In order to use this inference procedure all facts must be in conjunctive-normal-form (described in the previous section). An example of the basic rule of inference for resolution is given below:

$$\frac{\begin{array}{ccc} \alpha & \vee & \beta \\ \neg\alpha & \vee & \gamma \end{array}}{\begin{array}{ccc} \beta & \vee & \gamma \end{array}}$$

This rule states that if you know that the clauses on the top two lines are true, then you can conclude that the clause on the bottom line is also true. The resolution procedure matches a literal in one clause against the negation of that literal in the second clause. In this example the literal α in the first clause is matched with its negation $\neg\alpha$ in the second clause. If such a match can be found, then we can conclude that the clause consisting of the disjunction of the remaining literals from the matching clauses must also be true. In this example the remaining literals from the first clause are $\{\beta\}$, and those from the second clause are $\{\gamma\}$. Therefore, the clause $\beta \vee \gamma$ must also be true.

The matching of literals is based on the most general unification (MGU) of two expressions involving variables [44]. In general, two expressions unify if they are syntactically identical after the variables in each are replaced by some assignment (the same assignment is used for both expressions). For example, the literal (Father x Art) matches the literal (Father Bob y) if we replace the variable x with Bob, and the variable y with Art.

The following table defines a collection of facts for the connection example described at the start of this section:

C1: ¬(Conn x y) ∨ ¬(= a Val(x)) ∨ (= a Val(y))
C2: (Conn D74-in1 M1-in2)
C3: (= 1 Val(D74-in1))

A trace of the rule applications that conclude that the second input of the first multiplier is also 1 is given below. The labels on the right indicate the bindings of the variables in the resolution step, and the two clauses that were resolved to produce the clause on that line.

D1:	¬(= a Val(D74-in1)) ∨ (= a Val(M1-in2))	x=D74-in1 y=M1-in2	C1	C2
D2:	(= 1 Val(M1-in2))	a=1	D1	C3

This example illustrated how we can use the resolution inference procedure to conclude additional facts from an initial set of facts. There are many different inference procedures based on resolution. Another useful proof procedure is resolution refutation which can be used to *prove* facts. Similar to resolution, resolution refutation is complete and sound.

Using resolution refutation we would like to prove if a collection of facts T logically implies the goal p, i.e., $T \vdash p$. If $T \cup \neg p \vdash nil$, then it must be the case that $T \vdash p$, since a literal and its negation resolve to nil ($p \neg p \vdash nil$). In resolution refutation, thus, the original goal is negated and resolved with the original set of facts. If it is possible to resolve these facts to nil, then the original goal must be logically implied by the original set of facts.

The following table illustrates the use of resolution refutation to prove that the second input of the first multiplier must be 1 for the previous example. The original goal is (= u Val(M1-in2)), i.e., we have to prove that the value of the port M1-in2 is u (a variable). The result of the inference procedure returns a binding for the variables which makes the goal true. The trace of the proof procedure includes the bindings for the variables, followed by the two clauses that were unified to produce the clause on that line. The first clause is the negation of the goal.

E1:	¬(= u Val(M1-in2))			
E2:	¬(Conn x M1-in2) ∨ ¬(=u Val(x))	y=M1-in2 a=u	E1	C1
E3:	¬(= u Val(D74-in1))	x=D74-in1	E2	C2
E4:	*nil*	u=1	E3	C3

The variable binding in the last step binds u to 1. Thus, from the original set of facts we have proved that the goal (= u Val(M1-in2)) is true if we bind u to 1. In general, there can be many different bindings for the variables in the original goal that make it true, and the resolution refutation inference procedure can find any one, or all of them.

The first example illustrated how it is possible to deduce additional facts from an original set of facts using the resolution inference procedure.

Similarly, the previous example showed how it is possible to prove a goal from an initial set of facts using resolution refutation. We would also like to be able to ask *how* to achieve a goal given an initial set of facts. For example, how to control the second input of the first multiplier to 1. Given the original set of facts, it should be possible to deduce that the first input of D74 must be controlled to 1.

Resolution residue [20] is a general inference procedure for planning. Given an initial set of facts, and a specification of the primitive facts that can be assumed, the inference procedure returns a collection of assumable facts which achieve the goal. The original goal is negated and resolved with the original set of facts, similar to resolution refutation. The resolution residue inference procedure terminates when a clause with all assumable literals is derived (unlike resolution refutation which terminates when a null clause is derived). Similar to resolution, resolution residue is complete and sound.

The following table defines a collection of facts for the connection example described earlier. This table has one additional fact on the last line which asserts that we can assume any value for the port D74-in1.

F1:	¬(Conn x y) ∨ ¬(= a Val(x)) ∨ (= a Val(y))
F2:	(Conn D74-in1 M1-in2)
F3:	(Assumable ¬(= x Val(D74-in1)))

The following table presents a trace of the planning procedure for resolution residue to control the second input of the first multiplier to 1. The original goal is (= 1 Val(M1-in2)), and the first clause is the negation of this goal.

G1:	¬(= 1 Val(M1-in2))			
G2:	¬(Conn x M1-in2) ∨ ¬(=1 Val(x))	y=M1-in2 a=1	F1	G1
G3:	¬(= 1 Val(D74-in1))	x=D74-in1	G2	F2

Since the literal in the clause G3 is assumable, the resolution residue inference procedure terminates with the answer that the first input of D74 must be controlled to 1 to achieve the goal. In general there can be many ways to achieve a goal, and resolution residue is guaranteed to find any one, or any number of them.

Describing facts declaratively in predicate calculus has the advantage that the same collection of facts can be used for a variety of purposes, as illustrated above for deducing additional facts, proving facts and planning. The remainder of this section will show how the design description for the device D74, which was presented earlier, can be used for simulation, diagnosis, and test generation. In addition, we will briefly describe the importance of controlling the inference procedure in order to increase efficiency.

3.4.1 Simulation

Simulation involves propagating values from the inputs of components to their outputs. For example, if an input of an *and* gate is 0, then its output must also be 0. It is possible to simulate a design by adding additional facts that define the values of the input ports, and using the resolution inference procedure to deduce additional facts that are the consequences of the values at the inputs of the design.

For example, the following facts define the inputs of the device D74 to be 1, 3 and 2.

```
H1:  (= 1 Val(D74-in1))
H2:  (= 3 Val(D74-in2))
H3:  (= 2 Val(D74-in3))
```

By adding these facts to the original design facts A0 through A17, and B1 through B3, we can deduce that the outputs of D74 must be 5 and 8. A trace of the deductive procedure for simulating the design with these inputs is given in Table 3.7.

These resolution steps have deduced that the value of the first output port of D74 is 5 (in addition to deducing the values of the inputs and outputs of the multipliers and the first adder). Similarly, it is possible to deduce that the value of the other output port of D74 is 8. A more detailed discussion of simulation can be found in [52] for the *MARS* simulator.

3.4.2 Diagnosis

Diagnosis involves identifying a set of components in a design that can account for a fault. A fault corresponds to an erroneous output for some inputs. For example, with the inputs 1, 3, and 2, the first output of D74 should be 5. However, there is a fault in the device if the output is actually 2. It is possible to diagnose a fault by using the resolution residue inference procedure, after adding additional facts that specify the types of possible failures.

I1:	¬(Conn D74-in1 y) ∨ (= 1 Val(y))	H1	B3
I2:	(= 1 Val(M1-in2))	I1	A6
I3:	(= 1 Val(M2-in1))	I1	A7
I4:	¬(Conn D74-in2 y) ∨ (= 3 Val(y))	H2	B3
I5:	(= 3 Val(M1-in2))	I4	A8
I6:	(= 3 Val(M2-in1))	I4	A9
I7:	¬(Conn D74-in3 y) ∨ (= 2 Val(y))	H3	B3
I8:	(= 2 Val(M1-in2))	I7	A10
I9:	(= 2 Val(M2-in1))	I7	A11
I10:	¬(Mult x M1-in1 i2 o) ∨ ¬(= b Val(i2)) ∨ ¬(= c (× 3 b)) ∨ (= c Val(o))	I5	B1
I11:	¬(= b Val(M1-in2)) ∨ ¬(= c (× 3 b)) ∨ (= c Val(M1-out))	I10	A1
I12:	¬(= c (× 3 1)) ∨ (= c Val(M1-out))	I11	I2
I13:	(= 3 Val(M1-out))	×	
I14:	¬(Mult x M2-in1 i2 o) ∨ ¬(= b Val(i2)) ∨ ¬(= c (× 1 b)) ∨ (= c Val(o))	I3	B1
I15:	¬(= b Val(M2-in2)) ∨ ¬(= c (× 1 b)) ∨ (= c Val(M2-out))	I14	A2
I16:	¬(= c (× 1 2)) ∨ (= c Val(M2-out))	I15	I8
I17:	(= 2 Val(M2-out))	×	
I18:	¬(Conn M1-out y) ∨ (= 3 Val(y))	I13	B3
I19:	(= 3 Val(A1-in1))	I18	A12
I20:	¬(Conn M2-out y) ∨ (= 2 Val(y))	I17	B3
I21:	(= 2 Val(A1-in2))	I20	A13
I22:	¬(Adder x A1-in1 i2 o) ∨ ¬(= b Val(i2)) ∨ ¬(= c (+ 3 b)) ∨ (= c Val(o))	I19	B2
I23:	¬(= b Val(A1-in2)) ∨ ¬(= c (+ 3 b)) ∨ (= c Val(A1-out))	I22	A4
I24:	¬(= c (+ 3 2)) ∨ (= c Val(A1-out))	I23	I21
I25:	(= 5 Val(A1-out))	+	
I26:	¬(Conn A1-out y) ∨ (= 5 Val(y))	I25	B3
I27:	(= 5 Val(D74-out1))	I26	A16

Table 3.7: A trace of resolution for simulating a design.

In diagnosing a design we must remove the set of facts which define the type of each component for consistency, since at least one of these components must be failing, and therefore it no longer satisfies its specified behavior. For this design we must remove the facts A0 through A5.

The source of a failure can be a component or a connection. For simplicity, we will assume that connections never fail. The following facts assert that it is only possible for multipliers or adders to fail.

```
J1:  (Assumable ¬(Mult x i1 i2 o))
J2:  (Assumable ¬(Adder x i1 i2 o))
```

The symptom for this failure is that the first output of D74 should have been 5. Using the original set of design facts A6 through A17, B1 through B3, the initial inputs H1 through H3, and the failure assumptions J1 and J2, we identify a set of components that could have caused the fault using the resolution residue inference procedure.

A trace of the diagnosis process is given in Table 3.8. The first clause in this table is the negation of the original symptom.

Since all the literals in the last clause (K25) are directly assumable, we can terminate the resolution residue inference procedure. These inference steps have deduced that one of M1, M2, or A1 is not working. None of the other parts, M3 or A2 can be responsible for the failure at the first output of D74. A more detailed discussion of diagnosis can be found in [23] for the *DART* diagnostitian.

3.4.3 Test Generation

The previous subsection showed how to use the resolution residue inference procedure to identify a collection of components whose failure can account for a fault. We can generate tests to help discriminate between the set of suspects. Each test checks some behavior of one or more suspects. If the outputs of the device are different from the values specified by the test, then the failure must be restricted to one of the suspects being tested. By generating and executing a collection of tests on the device it is possible to identify a smaller set of suspects that are responsible for the fault.

Generating tests for a suspect requires controlling its inputs to some value and observing its output. Since we cannot control the inputs and outputs of the suspects directly, we must control the inputs of the device

K1:	¬(= 5 Val(D74-out1))		
K2:	¬(Conn x D74-out1) ∨ ¬(= 5 Val(x))	K1	B3
K3:	¬(= 5 Val(A1-out))	K2	A16
K4:	¬(Adder A1 A1-in1 A1-in2 A1-out) ∨ ¬(= a Val(A1-in1)) ∨ ¬(= b Val(A1-in2)) ∨ ¬(= 5 (+ a b))	K3	B2
K5:	¬(Adder A1 ...) ∨ ¬(= 3 Val(A1-in1)) ∨ ¬(= 2 Val(A1-in2))	+	
K6:	¬(Adder A1 ...) ∨ ¬(Conn x A1-in1) ∨ ¬(= 3 Val(x)) ∨ ¬(= 2 Val(A1-in2))	K5	B3
K7:	¬(Adder A1 ...) ∨ ¬(= 3 Val(M1-out)) ∨ ¬(= 2 Val(A1-in2))	K6	A12
K8:	¬(Adder A1 ...) ∨ ¬(Mult M1 ...) ∨ ¬(= a Val(M1-in1)) ∨ ¬(= b Val(M1-in2)) ∨ ¬(= 3 (× a b)) ∨ ¬(= 2 Val(A1-in2))	K7	B1
K9:	¬(Adder A1 ...) ∨ ¬(Mult M1 ...) ∨ ¬(= 3 Val(M1-in1)) ∨ ¬(= 1 Val(M1-in2)) ∨ ¬(= 2 Val(A1-in2))	×	
K10:	¬(Adder A1 ...) ∨ ¬(Mult M1 ...) ∨ ¬(Conn x M1-in1) ∨ ¬(= 3 Val(x)) ∨¬(= 1 Val(M1-in2)) ∨ ¬(= 2 Val(A1-in2))	K9	B3
K11:	¬(Adder A1 ...) ∨ ¬(Mult M1 ...) ∨ ¬(= 3 Val(D74-in2) ∨ ¬(= 1 Val(M1-in1)) ∨ ¬(= 2 Val(A1-in2)	K10	A8
K12:	¬(Adder A1 ...) ∨ ¬(Mult M1 ...) ∨ ¬(= 1 Val(M1-in2)) ∨ ¬(= 2 Val(A1-in2)	K11	H2
K13:	¬(Adder A1 ...) ∨ ¬(Mult M1 ...) ∨ ¬(Conn x M1-in2) ∨ ¬(= 1 Val(x)) ∨¬(= 2 Val(A1-in2)	K12	B3
K14:	¬(Adder A1 ...) ∨ ¬(Mult M1 ...) ∨ ¬(= 1 Val(D74-in1)) ∨ ¬(= 2 Val(A1-in2)	K13	A6
K15:	¬(Adder A1 ...) ∨ ¬(Mult M1 ...) ∨ ¬(= 2 Val(A1-in2))	K14	H1
...			
K25:	¬(Adder A1 A1-in1 A1-in2 A1-out) ∨ ¬(Mult M1 M1-in1 M1-in2 M1-out) ∨ ¬(Mult M2 M2-in1 M2-in2 M2-out)	K24	H3

Table 3.8: A trace of resolution residue for diagnosing faults.

and observe an output of the device. The inputs of the device must be controlled to control the inputs of the suspect being tested, and to propagate the output of the suspect being tested to an output of the device.

Given a test for a suspect, we can use the resolution residue inference procedure to deduce the values for the inputs of the device, and the value to observe at an output of the device. The inputs to the inference procedure are the design facts, the test to achieve, and the assumable facts for the planning procedure.

We can generate tests to discriminate between the set of suspects in the previous example. For example, we can generate a test to check the behavior of the second multiplier M2. The set of design facts for this example are A0 through A17 and B1 through B3. One test for this multiplier requires controlling its inputs to the value 1 and 2, and observing the value 2 at its output. In addition, assume that we can only control the inputs of D74, and only observe its outputs. The following table defines these assumable facts.

L1:	(Assumable ¬(= x Val(D74-in1)))
L2:	(Assumable ¬(= x Val(D74-in2)))
L3:	(Assumable ¬(= x Val(D74-in3)))
L4:	(Assumable (= x Val(D74-out1)))
L5:	(Assumable (= x Val(D74-out2)))

A trace of the deductive procedure for achieving the test for M2 is given in Table 3.9. The first clause in this table is the local test for M2.

Since all the literals in the last clause (M21) are directly assumable, we can terminate the resolution residue inference procedure. These inference steps have deduced that one way to control the inputs of M2 to 1 and 2, and to observe a 2 at its output, is to control the inputs of D74 to 1, 3, and 2, and to observe the value 8 at the second output of D74. A more detailed discussion of the test generation process can be found in the next chapter.

3.4.4 Control

In general, at each step of an inference procedure there are many different choices in deciding which clauses to resolve. In the previous examples we chose the most direct sequence of resolution operations to solve a goal. Without the guided application of these resolution operations the inference procedure can get extremely inefficient.

M1:	¬(= 1 Val(M2-in1)) ∨ ¬(= 2 Val(M2-in2)) ∨ (= 2 Val(M2-out))		
M2:	¬(Conn x M2-in1) ∨ ¬(= 1 Val(x)) ∨ ¬(= 2 Val(M2-in2)) ∨ (= 2 Val(M2-out))	M1	B3
M3:	¬(= 1 Val(D74-in1)) ∨ ¬(= 2 Val(M2-in2)) ∨ (= 2 Val(M2-out))	M2	A7
M4:	¬(= 1 Val(D74-in1)) ∨ ¬(Conn x M2-in2) ∨ ¬(= 2 Val(x)) ∨ (= 2 Val(M2-out))	M3	B3
M5:	¬(= 1 Val(D74-in1)) ∨ ¬(= 2 Val(D74-in3)) ∨ (= 2 Val(M2-out))	M4	A10
M6:	¬(= 1 Val(D74-in1)) ∨ ¬(= 2 Val(D74-in3)) ∨ ¬(Conn M2-out y) ∨ (= 2 Val(y))	M5	B3
M7:	¬(= 1 Val(D74-in1)) ∨ ¬(= 2 Val(D74-in3)) ∨ (= 2 Val(A2-in1))	M6	A14
M8:	¬(= 1 Val(D74-in1)) ∨ ¬(= 2 Val(D74-in3)) ∨ ¬(Adder x A2-in1 i2 o) ∨¬(= b Val(i2)) ∨ ¬(= c (+ 2 b)) ∨ (= c Val(o))	M7	B2
M9:	¬(= 1 Val(D74-in1)) ∨ ¬(= 2 Val(D74-in3)) ∨ ¬(= b Val(A2-in2)) ∨¬(= c (+ 2 b)) ∨ (= c Val(A2-out))	M8	A5
M10:	¬(= 1 Val(D74-in1)) ∨ ¬(= 2 Val(D74-in3)) ∨ ¬(= 6 Val(A2-in2)) ∨(= 8 Val(A2-out))	+	
M11:	¬(= 1 Val(D74-in1)) ∨ ¬(= 2 Val(D74-in3)) ∨ ¬(Conn x A2-in2) ∨¬(= 6 Val(x)) ∨ (= 8 Val(A2-out))	M10	B3
M12:	¬(= 1 Val(D74-in1)) ∨ ¬(= 2 Val(D74-in3)) ∨ ¬(= 6 Val(M3-out)) ∨(= 8 Val(A2-out))	M11	A15
M13:	¬(= 1 Val(D74-in1)) ∨ ¬(= 2 Val(D74-in3)) ∨ ¬(Mult x i1 i2 M3-out) ∨¬(= a Val(i1)) ∨ ¬(b Val(i2)) ∨ ¬(= 6 (× a b)) ∨ (= 8 Val(A2-out))	M12	B1
M14:	¬(= 1 Val(D74-in1)) ∨ ¬(= 2 Val(D74-in3)) ∨ ¬(= a Val(M3-in1)) ∨¬(b Val(M3-in2)) ∨ ¬(= 6 (× a b)) ∨ (= 8 Val(A2-out))	M13	A3
M15:	¬(= 1 Val(D74-in1)) ∨ ¬(= 2 Val(D74-in3)) ∨ ¬(= 2 Val(M3-in1)) ∨¬(3 Val(M3-in2)) ∨ (= 8 Val(A2-out))	×	
M16:	¬(= 1 Val(D74-in1)) ∨ ¬(= 2 Val(D74-in3)) ∨ ¬(Conn x M3-in1) ∨¬(= 2 Val(x)) ∨ ¬(3 Val(M3-in2)) ∨ (= 8 Val(A2-out))	M15	B3
M17:	¬(= 1 Val(D74-in1)) ∨ ¬(= 2 Val(D74-in3)) ∨ ¬(3 Val(M3-in2)) ∨(= 8 Val(A2-out))	M16	A11
M18:	¬(= 1 Val(D74-in1)) ∨ ¬(= 2 Val(D74-in3)) ∨ ¬(Conn x M3-in2) ∨(= 3 Val(x)) ∨ (= 8 Val(A2-out))	M17	B3
M19:	¬(= 1 Val(D74-in1)) ∨ ¬(= 2 Val(D74-in3)) ∨ ¬(= 3 Val(D74-in2)) ∨(= 8 Val(A2-out))	M18	A9
M20:	¬(= 1 Val(D74-in1)) ∨ ¬(= 2 Val(D74-in3)) ∨ ¬(= 3 Val(D74-in2)) ∨¬(Conn A2-out y) ∨ (= 8 Val(y))	M19	B3
M21:	¬(= 1 Val(D74-in1)) ∨ ¬(= 2 Val(D74-in3)) ∨ ¬(= 3 Val(D74-in2)) ∨(= 8 Val(D74-out2))	M20	A17

Table 3.9: A trace of resolution residue for generating a test.

Automated deduction using resolution is a combinatoric process. It is possible to conclude all the facts in the deductive closure of the initial set of facts under logical implication. For a given problem, most of these facts will be irrelevant, and any resources used in deriving them constitute wasted effort.

In order to increase the efficiency of the reasoning process it is important to focus the resolving of clauses to solve the problem at hand. For simulation, diagnosis, and test generation it is possible to use the *linear-input-form* and *set-of-support* control strategy to increase the efficiency [44]. These control strategies increase the efficiency of the reasoning procedure without restricting the types of designs that can be simulated, diagnosed, or tested.

The set-of-support control strategy always chooses at least one of the clauses to be resolved from the set of clauses derived from the negation of the original goal (including the negation of the original goal). This control strategy focuses the reasoning procedure on the problem at hand without affecting the completeness of the inference procedure. The linear-input-form control strategy always chooses one of the clauses to be resolved from the initial set of facts (in reasoning about designs, this is the set of facts describing a design). This control strategy is more efficient than set-of-support by itself. However, the reasoning procedure is no longer complete. Fortunately, the inference procedure is still complete for simulation, diagnosis, and test generation, although it is logically incomplete.

Even with the combination of these two control strategies there can be many choices in deciding which clauses to resolve. Traditional approaches use a static criteria to make this selection. For example, the language PROLOG scans the knowledge base from the start, and uses the first clause with a literal that matches a literal in the most recently derived clause. Unfortunately, the selection criteria based on the static ordering of clauses is not efficient for all tasks. For example, in reasoning from the inputs of a device to the outputs, a different ordering might be required than when reasoning from the outputs to the inputs.

The solution to this problem is to use a control strategy that can be modified dynamically based on the current situation. Such a flexible control strategy is provided by MRS [49], which allows the user to specify the control of the problem solving process, in addition to defining the problem area. The system provides a vocabulary for specifying that the clauses should be resolved in a situation, independent of the ordering of these clauses in the knowledge base.

There is an overhead associated with the meta-level effort of deciding which clauses to resolve. For simple problems, such as simulation, the increase in performance is not worth the overhead of the flexible control of the inference procedure. However, for complex tasks, such as diagnosis and test generation, the improvement in performance is definitely worth the overhead. A more detailed discussion of the control strategies for test generation can be found in the next chapter.

3.5 Utility of General Representation and Reasoning

In this section we will describe the advantages of using general methods based on logic for representing and reasoning about designs, and also describe some of the disadvantages.

3.5.1 Advantages

As a representation language predicate calculus is task independent, since the information is represented *declaratively*. That is, there is a precise definition of the interaction between the parts of a description, and in addition, these interactions are only based on pattern matching parts of expressions via unification. The same description of a design can be used for a collection of tasks, e.g., simulation, diagnosis, and test generation, as illustrated in the previous section.

Predicate calculus is a *general* representation language for describing designs. We are free to choose the universe of discourse U, the constant symbols <const>, and the interpretation function I. By the appropriate definition of these we can define a model for an arbitrary design $D = < O, F, R >$. The universe of discourse U is the same as the set of objects in the design O. For each such object $\alpha \in O$ we must define an <obj-const> symbol alpha such that $I(\text{alpha}) = \alpha$. For each function $f \in F$ $(f : O^n \rightarrow O)$ we must define a <func-const> f such that $I(\text{f}) \subseteq U^{n+1}$, where the first n positions in a tuple correspond to the arguments of the function, and the $n+1$ position corresponds to the result of the function. Similarly, for each relation $r \in R$ $(r : O^n \rightarrow \{true, false\})$ we must define a <rel-const> r such that $I(\text{r}) \subseteq U^n$.

The generality of the language permits describing arbitrary devices, e.g., digital circuits, and mechanical devices. In addition, the generality of the language permits describing a single design at a collection of abstraction levels in a uniform framework, e.g., describing a digital circuit at the gate level and the register transfer level. Having a uniform framework simplifies the task of keeping the specifications of a design at the different abstraction levels consistent with each other (as opposed to having distinct languages for each level). The descriptions of a device at the different abstraction levels can also be related to each other, thus permiting shifting between levels of abstractions in reasoning about a device. This flexibility can be exploited to increase efficiency by always reasoning at the most abstract level possible.

Another advantage of the generality of the language is that it permits specifying partial descriptions of a device. For example, the design description presented earlier for the device D74 is incomplete since it is not refined to a level where it is possible to realize the device (e.g., down to the gate level). In the absence of a complete design it is still possible to reason about the device and draw useful conclusions, e.g., we can localize a fault in the device to the second multiplier, without identifying the part of this multiplier that is responsible for the fault. A more realistic example is in modeling a complicated device, such as the Intel 8086 microprocessor. We can encode a partial description of this device in predicate calculus by capturing the high level black box behavior specified in its technical manual. This partial description can be used to reason about a larger design of which the microprocessor is a part. It would be extremely difficult to encode such a partial description using an abstraction specific language, e.g., boolean algebra.

An important advantage of describing designs in predicate calculus is that it has a well defined syntax and semantics, and a complete and valid proof procedure. In addition, the ability to quantify propositions enables describing continuous behavior, and partial and uncertain information.

The ability to universally quantify propositions allows the description of behavior which is continuous in time, and relative to any given time. For example, the following facts describe the behavior of an *or* gate with a delay of 5 time units:

```
N1:  (= [1 t] Val(Or-in1)) ⇒ (= [1 +(t 5)] Val(Or-out))
N2:  (= [0 t] Val(Or-in1)) ∧ (= [x t] Val(Or-in2)) ⇒ (= [x +(t 5)] Val(Or-out))
```

The first fact asserts that if the first input of the *or* gate is 1 at any time t, then its output will also be 1 after a delay of 5 time units. Similarly, the second fact asserts that if the first input is 0 at some time t, and the the second input is x at the same time, then the output will also be x after a delay of 5 time units.

Universally quantifying variables permits collapsing descriptions where the specific value of a term is not important (instead of enumerating all possible values for a variable). For example, if the first input of the or-gate is 0, then the output is the same as the second input after a delay of 5 time units (this is true for any value at the second input).

The ability to quantify propositions existentially permits specifying partial information (which may be made more specific as a design evolves). For example, we can specify descriptions for a class of objects, without naming the objects directly. This is the case when defining prototypes in a library of designs, where the individual instances of this prototype are not known in advance. A typical prototype definition for a *nand* gate is given below, where each *nand* gate consists of an *and* gate connected to an inverter.

```
∀w∀a∀b∀c(Nandgate w a b c)  ⇒  ∃x∃y∃d∃e(Andgate x a b d) ∧
                                (Inverter y e c) ∧ (Conn d e) ∧
                                (= Val(c) F-nand(Val(a) Val(b)))
```

The proposition (Nandgate Foo In1 In2 Out) asserts that Foo is a *nand* gate, whose input ports are In1 and In2, and whose output port is Out (similarly for (Andgate x a b c) and (Inverter x a b)). These definitions make use of the function F-nand which is defined below:

```
∀x∀t   (= F-nand([0 t] [x t]) [1 +(t 5)]) ∧
∀x∀t   (= F-nand([1 t] [x t]) [not(x) +(t 5)]) ∧
       (= Not(1) 0) ∧ (= Not(0) 1)
```

Another advantage of representing designs in predicate calculus is that the language permits representing and reasoning with uncertain information, e.g., disjunctive propositions. This capability is essential for many tasks in reasoning about a device, e.g., diagnosis. In diagnosing a device we incrementally accumulate pieces of information by executing tests. Each piece of information usually does not identify the faulty component, instead it usually identifies a collection of components, one or more of which is responsible for the fault. By combining different pieces of uncertain information it is possible to identify the single component responsible for the fault. For example, if we know the following facts:

```
(Notworking And1)∨(Notworking And2)∧(Subpart And1 Sn7408)∧
(Subpart And2 Sn7408) ⇒ (Notworking Sn7408)
```

That is, if you know that one of And1 or And2 is not working, and that both And1 and And2 are a part of Sn7408, then you can conclude that

Sn7408 itself must not be working (a component is not working if any of its parts are not working).

3.5.2 Disadvantages

The disadvantage of using general methods based on logic for representing and reasoning about designs is the inefficiency for specialized and restricted domains. The generality of these methods has an overhead associated with it, e.g., the cost of the unification procedure, which is computationally very expensive.

For small designs the overhead associated with this generality will be unacceptable, since it may be inappropriate to describe such designs at a collection of abstraction levels. This eliminates the principal advantage of the general methods, i.e., reasoning at abstract levels to increase efficiency. It is more appropriate instead to increase the efficiency by restricting the generality of the reasoning procedure for such applications, e.g., using specialized inference procedures for boolean logic for designs described at the gate level.

Although predicate calculus is a useful language for representing and reasoning about designs it is not the most appropriate language for designers to use directly. However, this does not preclude translating a language that is easier to use by designers (e.g., graphical languages for describing structure and behavior) into predicate calculus.

Chapter 4

Test Generation

The previous chapter described the utility of using general methods based on logic for representing and reasoning about designs, and illustrated the reasoning procedures for simulation, diagnosis, and test generation. In this chapter we will present a more detailed discussion of the test generation task. We will concentrate on the control issues for increasing the efficiency of the reasoning process, and ignore the details of the representation and inference procedures which were described in the previous chapter.

As a task, test generation is interesting since its cost is exponential in the size of a design, and any effective test generation strategy must manage this complexity for large designs. From a pragmatic standpoint test generation is important since it is an integral part of manufacturing a device. With the increase in the complexity of devices that can be manufactured it is important to find a solution for this task.

In this chapter we will first define the test generation task. We will also briefly describe the work done in the past to solve this problem, and show the limitations of these approaches. We will next describe the SATURN test generation system which exploits the methodology proposed in this thesis to manage the complexity of test generation (i.e., reasoning with abstract design formulations). In describing the SATURN test generation system we will first present the overall architecture, and then describe the control strategies for increasing the efficiency of the reasoning procedures via an example. Finally, we will present some empirical results that demonstrate the utility of these control strategies, and the utility of reasoning with high level design descriptions.

4.1 Task Definition

Test generation can involve testing a design, or testing a device. In the former case we are not sure if the different parts of a design are mutually consistent, and in the latter case we are not sure if a device is functioning correctly. For both these cases, test generation involves generating a collection of tests, which when executed on the design, or device, check its functionality. The result of the test generation process is a collection of tests, where each test specifies the values for some inputs, and the values to observe at some outputs for these inputs. In this thesis we are focusing on testing the functionality of a design/device, and are not addressing testing other properties, e.g., power consumption, and the steady state voltage and current parameters.

Usually designs evolve incrementally over time, and are refined using a combination of top-down and bottom-up strategies [9]. In such an environment we would like to verify each incremental refinement of a design before proceeding further. This involves generating tests that verify that the high level behavior of a component is consistent with the composition of the low-level behaviors of its subparts. Verifying the consistency between the descriptions at the two abstraction levels requires generating tests that exercise the behavior of the top-level function completely, and comparing the outputs of the two descriptions for these inputs. This corresponds to testing designs via simulation, and it is not always the most efficient means of verifying the equivalence between two descriptions. It is more efficient to compare the two symbolic design descriptions algebraicly, without requiring simulation, as is done in the design verification system VERIFY [4].

In this thesis we are only going to consider the latter use of test generation for testing a device. In this case the design is assumed to be correct, however, the manufacturing process which realizes the physical device from the design specification is assumed to be imperfect. The imperfections in the realization process mandate that every device be tested after it is manufactured. Since the device is a real physical entity, its functionality must be verified experimentally be executing a collection of tests. The goal of test generation is to come up with a sequence of tests, such that if the device satisfies these tests it is guaranteed to be consistent with its design. This goal must be satisfied subject to certain constraints, e.g., minimizing the length of the test vectors. In practice, it is impractical to generate the minimal set of test vectors to test a device, and a small set is acceptable.

In the absence of any information about the possible manufacturing failures, we must test the behavior of the device for all input/state combinations. By knowing the possible failures we can reduce the number of tests by only generating tests for the possible failures, and combining tests with compatible inputs (tests which agree on the values of all inputs). For example, we can reduce the number of tests for digital circuits by assuming that it is only possible for the inputs and outputs of boolean gates to fail by being stuck-at 1 or 0. In addition, the number of tests can be reduced dramatically by making further assumptions about the nature of the possible failures, e.g., the *single fault* and *non-intermittency* assumption. The single fault assumption states that only one of the possible failures can occur at any given time. Similarly, the non-intermittency assumption asserts that failures are permanent, and not intermittent over time.

Assuming that the possible manufacturing failures are stuck-at 1, or stuck-at 0, faults at the boolean gate level (including the single fault and non-intermittency assumption), the test generation process must generate tests to check each possible fault for all the boolean gates in the device. For example, if *xor*51 is an exclusive-or gate in the device, this includes checking for stuck-at 1, and stuck-at 0, faults at both inputs and the output. A test for checking if the output is stuck-at 0 requires controlling the inputs so that the output should be 1 if the device is working, and checking the actual output for these inputs. For example, if the inputs of an exclusive-or gate are 1 and 0 and the output is 1, then the output cannot be stuck-at 0. However, if the output is 0, one possible failure is that the output of the gate is stuck-at 0 (it is also possible for the first input to be stuck-at 0, or the second input to be stuck-at 1). In testing a device we are only interested in checking if it is functioning or not. Due to random manufacturing defects a certain percentage of the devices will always fail, and we are not interested in isolating the causes of these failures.

The testing of individual failures is complicated by the fact that not all ports of the device are directly controllable or directly observable. In current practice usually only only a small fraction at the perimeter of the device are directly controllable and/or directly observable. In order to test a subpart of a device we must set the directly controllable inputs so that the part being tested has the correct inputs and its output is propagated to a directly observable port. The exact value propagated to a directly observable output is not important as long as it is a function of the output of the part being tested. That is, if the part being tested is functioning we get one value at the output, and if it is failing we get any *different* value at the same output.

The ideal design formulation for test generation would directly specify how to achieve a test, i.e., it would specify how to control the directly controllable inputs to control the inputs of a subpart being tested, and also specify how to control the directly controllable inputs to propagate the output of the subpart being tested to a directly observable output. Unfortunately, the size of such a design would be extremely large, since it would require specifying the value of every internal port as a function of the directly controllable inputs, and specifying the value of every directly observable output as a function of each internal port and the directly controllable inputs. A more acceptable tradeoff is to partition the design into a collection of components and propagate values through these components to achieve a test.

In propagating values through a design there are usually many local choices that achieve a subgoal. For example, there are three input combinations that achieve the subgoal of controlling the output of an *and* gate to *false*. Similar choices exist in propagating the result of a test to a directly observable output, e.g., in selecting one of the branches of a fanout node, or one of the outputs of a module. Test generation, therefore, involves search. Unfortunately, not all paths in the search space are guaranteed to lead to a solution. This is due to the global interactions between the different local choices that are made in achieving a test. The interactions are caused by the reconvergent fanouts in a device, which correspond to the constraints that all branches of a fanout must have the same value. Reconvergent fanouts are quite common in devices, since fanouts correspond to an optimization where hardware is shared between different functions to minimize area, power and cost. Therefore, many paths in the search space for test generation can lead to dead-ends, and a successful test generation system must navigate through this search space to avoid as many of these failure paths as early as possible.

The key problem in testing is to generate a reasonably small set of vectors for testing a device. The execution of a single test is relatively inexpensive, and the number of vectors required to test a device is linear in the number of parts [25]. A naive test generation algorithm would enumerate all input/state combinations, however, the number of test vectors generated will be unacceptably large. The key problem in reducing the number of test vectors to test a device is the cost of test generation, which is non-linear, and sometimes exponential (in fact it is NP-complete [30]). This complexity is compounded for sequential circuits where we may have to unfold a copy of a finite state machine for each possible state. In practice, the most optimistic estimates show that the cost of test generation is proportional to the cube of the circuit size [25].

4.2 Previous Work

In this section we will briefly describe previous approaches to the test generation task, and describe their limitations for reasoning about complex devices.

Along one dimension we can classify the various test generation schemes by the types of devices they generate tests for. Highly specialized testing strategies have been developed for testing devices with regular and repetitive structures, e.g., for testing random access memories (RAMs) and programmed logic arrays (PLAs). These devices are densely packed, and consequently the failures are dependent on combinations of values of adjacent nodes in the device. Depending on the actual physical layout of the device, different test patterns can be generated which test these failure modes [36].

Alternate testing schemes are applicable for devices that have been designed following a specific design methodology to reduce the complexity of test generation. One such methodology consists of adding additional hardware in the device that adds a scan path through all the sequential circuits [38]. The scan path is a single long shift register that permits controling and observing the inputs and outputs of all combinational logic blocks in the device. The additional hardware transforms the problem of testing sequential circuits into the problem of testing a collection of combinational circuits. The increase in the controllability and observability of the device greatly simplifies the test generation task, however the length of the test vectors is proportional to the square of the number of components in the device (instead of being linear).

Another design methodology to simplify the test generation task is to add hardware which allows a device to test itself [12]. The tests are usually generated within the device using a pseudo-random number generator, and the response is checked using signature analysis (again, within the device). This approach has the disadvantage that it requires additional hardware, and a long sequence of random vectors to test a device. In addition, random vectors give poor (less than 80%) fault coverage for sequential circuits, and the testing logic can itself fail (i.e., you cannot distinguish a failure in the device from a failure in the testing

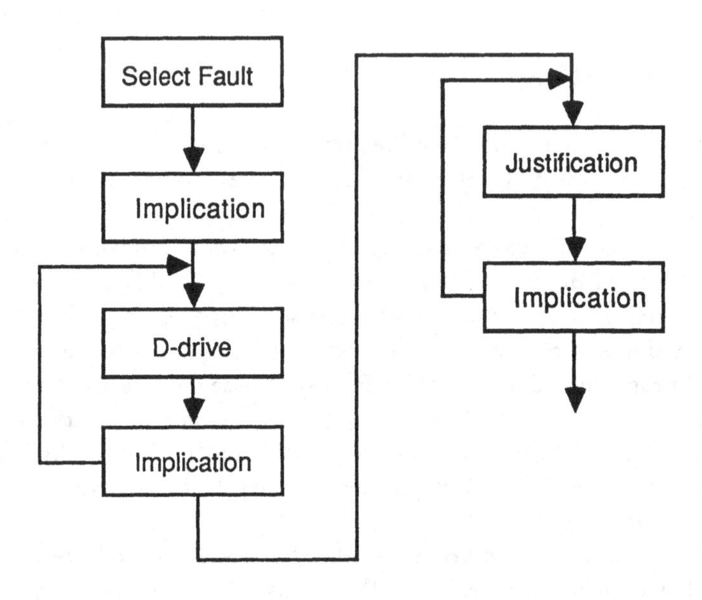

Figure 4.1: Control structure of the D-algorithm.

logic).

Other test generation schemes attempt to generate tests for devices that have been designed using an arbitrary methodology. One of the earliest, and most important of such algorithmic schemes is the D-algorithm [48]. This system accepts design descriptions at the boolean gate level and generates tests to check for stuck-at faults at the inputs and outputs of each gate in the design.

The overall control structure of the algorithm is given in Figure 4.1. The system first selects a fault to test, checking the consequences of this test through the design (implication). It next propagates the output of the device being tested to one of the directly observable outputs (d-drive), checking the consequences of each step. In the last stage the system propagates the inputs of the subparts that must be controlled back to the directly controllable inputs (justification), again checking the consequences of each step.

Each test for a subpart specifies the values for its inputs and the test output. The implication step propagates the consequences of a port having some value throughout the design. For example, if the first input of the device being tested is 0, then all other ports connected to this input must also be 0. If the input of another *and* gate is connected to this port its output must also be 0, and so on. The purpose of performing the implication step is to check if a local choice is globally inconsistent, thus requiring backtracking. For example, if one subgoal requires controlling the output of an *and* gate to 1, and the consequence of an earlier choice is that one of its inputs is 0, then this state is inconsistent. When an inconsistency is detected the D-algorithm undoes the latest choice, selects a new alternative, and checks the consequence of this alternative via implication.

The output of a device being tested is represented by the symbol d or $\bar{\text{d}}$ (called discrepancy signals). The symbol d indicates that the output will be 1 if the fault is absent, and will be 0 if the fault is present, and vice-versa for the symbol $\bar{\text{d}}$. The d-drive stage involves propagating the symbol d/$\bar{\text{d}}$ at the output of the device being tested to one of the directly observable outputs. Due to fanouts it is possible to propagate this signal along different paths in the design. Each step of the d-drive requires selecting a discrepancy signal at the frontier, and propagating it one step closer to a directly observable output through another gate. Propagating discrepancy signals from an input to the output of a gate can flip the sign of the discrepancy signal. For example, if the input of an inverter is d then its output will be $\bar{\text{d}}$ (if the fault being tested is absent the output of this inverter will be 0, and if it is present the output of this inverter will be 1). This propagation can also require controlling the other inputs of the gate through which the discrepancy signal is being driven. For example, in order to propagate a discrepancy signal from an input of an *and* gate to its output, the other inputs must all be controlled to 1 (otherwise the output will always be 0, independent of the fault being tested). The consequences of the inputs that must be controlled is checked to see if they are globally consistent via implication.

The justification stage of the procedure propagates the inputs of the gates that must be controlled back to the directly controllable inputs. This includes the inputs of the device being tested, and the other inputs that must be controlled for the d-drive. At each step the goal of controlling the output of a gate to some value is replaced by the subgoals of controlling some inputs of this gate. Many choices can exist in controlling the output of a gate, and the consequences of each choice must be checked to see that they are globally consistent via implication.

The advantage of the D-algorithm is that it is a complete test generation procedure for stuck-at faults at the gate level. If a test for a part of a device exists, the algorithm is guaranteed to find it (eventually).

The principal disadvantage of the D-algorithm is that it reasons with a flat gate level description of a device. Quite often it is hard to provide such a description if there is a mismatch between the primitives of the device and the language, e.g., describing the charge stored on an nMOS transistor in terms of boolean gates. Another limitation of the D-algorithm is that it can only generate tests for stuck-at faults at the gate level. For example, it cannot generate tests that check for shorts between physically adjacent signals. The critical problem with reasoning at the gate level, however, is the sheer complexity of the number of gates through which information must be propagated. This problem is further compounded if the device has feed back paths. Although the D-algorithm has been extended to handle sequential circuits (by virtual replication of finite state machines), it is not very efficient for large devices. There have been attempts made to get around this efficiency problem by using massively parallel architectures, like the connection machine [34], for generating tests. However, throwing hardware at a problem that is exponential in the size of the task is not a realistic solution.

Another disadvantages of the D-algorithm is the limitation of its control structure. There is no way of controlling the reasoning procedure to make it choose the more promising alternatives first, e.g., selecting shorter paths back to the inputs first, and only examining the other paths if this leads to an inconsistency. In addition, when an inconsistency is detected, the system backtracks to the last choice point, instead of the source of the inconsistency (possibly an earlier choice), as is done with dependency directed backtracking [18]. The D-algorithm also repeatedly solves the same subgoals since it does not cache solutions to subgoals, e.g., figuring out how to control a port to the same value repeatedly, each time for a different test. In addition a large number of tests can be generated since the system does not perform fault collapsing [8]. For example, every path from a directly controllable input to a directly observable output in a given test checks for half the stuck-at faults for all the ports in this path.

Other algorithmic approaches attempt to solve test generation problem by generating tests from high-level descriptions of devices, where there are fewer components to reason with. These schemes are usually classified as *functional testing*. For example, in one such system

[35] a design is specified using a graph description language to define a state-transformation model of a device. The description of a device is separated into the control and data section, where the primitive graph operators are similar to those found in data-flow languages (forks, joins, merges, predicates, selectors, operators, etc.). In the absence of a low-level gate description the user must define the fault model for the nodes in the graph. The disadvantage of this approach is that it only provides a single abstraction level for describing designs, and the fault models for the graph nodes must be generated outside the system (e.g., using the D-algorithm).

Other test generation schemes attempt to manage the cost of generating tests for complex devices by making the test generation process interactive, i.e., by integrating the user in the test generation process. Examples of this are the SCIRTSS [28] and HITEST [47] test generation systems. In the SCIRTSS system, a design is specified as a state table, and the system verifies the transitions of the device from one state to another. Similar to functional testing, the user must specify the data values to use in checking these state transitions (otherwise a single random value is chosen). In addition, the user can specify parameters which act as heuristics to guide the search in test generation. If the system fails to achieve a test after 50 backtracking operations the user is asked to modify the search heuristics. The actual propagation of the tests within combinational logic blocks is achieved using the D-algorithm.

The HITEST system attempts to be more ambitious by trying to integrate *knowledge items* from an experienced human test programmer. These pieces of knowledge can capture how to test a given collection of components, e.g., how to test an adder made up of four full-adders. The test specification for components can be parameterized to specify a schema for testing a class of similar components, e.g., to test adders with varying number of input bits. In addition, the user can specify high-level plans for testing parts of a device. These plans provide intermediate islands in the search space, which constrain the overall test generation process. The actual generation of tests is accomplished using the PODEM [26] test generation procedure. If more than 9 backtracking operations are required in generating tests the system gives up and asks the user for help.

The PODEM algorithm corresponds to a generate and test procedure for achieving tests. Heuristics are employed to select the next input pattern to consider, by selecting the next input to constrain, and selecting the value to constrain it to. The consequences of each new input value are propagated through the design, to see if the desired test is achieved. In the worst case all input combinations must be explored. PODEM works best for combinational functions of exclusive-or gates with reconvergent fanout (typical of error correction and translation circuits in computers).

A feature of the existing test generation algorithms is that they trade-off completeness for efficiency. Attempts to model complex devices at a high-level, without relating the high-level descriptions to the low-level models (which have the appropriate granularity for selecting the fault models) improves the efficiency of test generation at the expense of completeness. For example, many of the high-level test generation schemes choose a single (or a small number) of values in checking the data paths of a device. Yet without knowing the fault model and detailed structure at the level of the possible faults, there is no rationale for assuming that the device has been completely tested.

Although many test generation systems claim to be hierarchical (e.g., the hierarchical D-algorithm), the hierarchy of design is only in structure and not behavior. That is, it is possible to impose a hierarchical structure on a design. However, behavior must still be defined in terms of boolean objects.

4.3 The Saturn Test Generation System

The Saturn test generation system integrates a collection of ideas from existing test generation systems. At its core, it is an algorithmic test generation system, similar to the D-algorithm, however, it is much more flexible. The main differences between Saturn and the D-algorithm are: it is possible to generate tests for designs specified at arbitrary abstraction levels in Saturn; the control structure of Saturn is more flexible since it permits selecting the most promising choices in the search space in a situation dependent manner; the Saturn system permits the user to specify the fault models (we are not restricted to stuck-at faults at the boolean level); the user can specify how to test a class of components; the Saturn test generation system minimizes computational effort by caching tests and solutions to subgoals in achieving tests; the Saturn test generation system also reduces the number of tests generated by collapsing tests. Similar to the D-algorithm, the Saturn test generation system is a complete procedure for generating tests.

In the reminder of this section we will describe the Saturn test generation algorithm in greater detail. We will first present an overview of the algorithm, and later illustrate this using as an example. We will next describe the control strategies for improving efficiency, and finally present some empirical results which demonstrate the utility of reasoning with reformulated designs.

4.3.1 Algorithm

In this subsection we will briefly describe the algorithm used in the Saturn test generation system.

The inputs to the Saturn system are the specifications of: the design, the directly controllable inputs and outputs, the fault models, the parameterized test clichés that specify local tests for a class of components, and the user specifications of what level of abstraction to test a component at. The result of the test generation process is a collection of tests that check all the possible faults in the design, or achieve the tests specified by the user in the clichés. The system also returns a list of the tests that could not be achieved.

The system first examines the topology of the circuit and computes the deductive cost estimates for controlling and observing the value of every port (separate estimates for controlling/observing). These costs estimate the number of inference steps required to control/observe the value of a port. These estimates are generated once and used repeatedly to select the most promising paths in the search space when propagating test values through the design.

After computing the deductive cost estimates the system starts the actual test generation process. The overall control scheme used by the test generator is to reason at the most abstract level possible in generating tests. This translates to examining the substructure of a component only when it is absolutely necessary to do so. In deciding how to test a component, and propagating values from its inputs to outputs (and vice-versa), the cost estimates guide the system to use the most abstract descriptions possible.

There are three sources of information that can be used for deciding how to test any module: its behavior specification, its substructure (the submodules and their interconnection), or a user supplied test sequence (a test cliché). For each module the user can specify which of these sources of information is to be used in testing it.

If the user specifies that a component is to be tested using a test cliché, then the substructure of the component need never be examined. Similarly, if the user specifies that a component is to be tested by using its behavior specification, the system generates tests that check all input/state combinations for the component. Finally, if the user specifies that a component is to be tested by testing its subparts, the system recursively generates tests for each subpart and combines these tests to define the tests for the parent component.

By allowing the user to specify how to test components it is possible to tradeoff between the quality of the tests and the time required to generate tests. For example, testing the behavior can be cheaper than generating tests from the substructure (if the number of inputs is much smaller than the number of subparts), however, it generates a larger set of test vectors (generating tests from the substructure allows fault collapsing which is not possible using the behavior). However, both alternatives check all possible faults.

If the user does not specify how to test a component, and there are no test clichés defined, then the system either tests the behavior of the component, or its subparts. If the fault models are defined at the current level, the system generates tests which test these faults without examining the substructure of a component. For example, in testing an *and* gate for a stuck-at 0 fault at the output it is only necessary to control the inputs so that the output is not 0. By examining the behavior description of the *and* gate we can deduce that both its inputs should be 1, and its output should also be 1. If the fault models are defined at a more refined level, the system chooses the alternative which it estimates will be cheaper (usually behavior, unless the number of input/state combinations is much larger than the number of subparts).

In deciding how to test a component by examining its substructure, the system recursively deduces how to test each subpart, and then propagates the tests for the subparts to the boundary of the parent (and not to the directly controllable inputs/outputs). The tests for the subparts are minimized by combining tests with compatible inputs, and are then abstracted to the next higher level (the level of the parent). Thus, we have generated the test specification for a component from the test specifications of its subparts.

The advantage of propagating the tests for the subparts up to the boundary of the parent (and not to the directly controllable inputs and outputs) is that we can develop test specifications for composite components. These specifications are generalized by the system so that they can be shared by all identical and similar components. For example, assume that the system has generated a test specification for a full-adder from the test specifications of its subparts. By caching this information it is never necessary for the system to examine the substructure of another

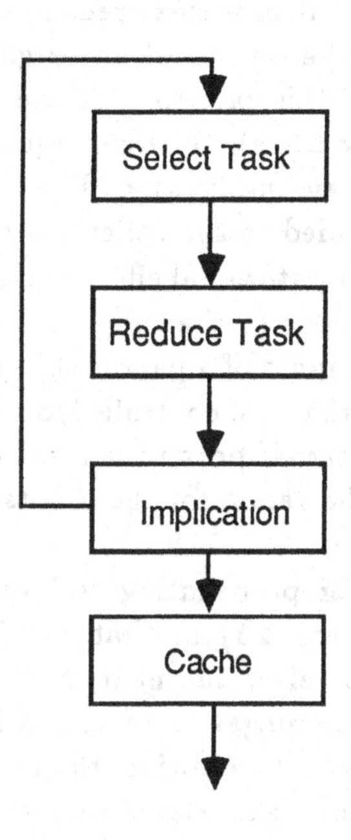

Figure 4.2: Control structure of Saturn for achieving tests.

full-adder (for the same fault models). The same test specifications can be used for all full-adders, in any design, with the same fault models.

The computational complexity of test generation arises from the choices which exist in propagating a test to the next higher level. Different choices can interact with each other due to reconvergent fanouts in a design. The system takes advantage of the deductive cost estimates to select more promising paths in propagating test values. Figure 4.2 shows the overall control algorithm for propagating values to the next higher level.

The system first selects a test for a subpart, and checks the consequences of this test by implication (similar to the D-algorithm). The consequences of a test are only propagated up to the hierarchy boundary of the parent (instead of the entire design). The system next dynamically chooses the cheapest task to achieve the test. Each task represents a different partial solution to the goal of propagating the original test to the next level. The selected task is next reduced one step, e.g., reducing the goal of controlling the output of an *or* gate to 1 to the goal of controlling the first input of this gate to 1, or the goal of controlling the second input to 1 (two new tasks). The consequences of the new tasks are checked for consistency via implication. If a task is inconsistent it is discarded, otherwise it is added to the collection of tasks.

In order to share the computational effort of one test with other tests the system caches the solutions to the subgoals of a test. This involves recording the values of the inputs of a parent that achieve a test, and the values of all internal ports that are controlled/observed by these values. If we have to control an internal port to the same value for a different test, then we can look up the values for the inputs of the parent directly instead of recomputing them.

The control structure for propagating test values in Saturn differs from the D-algorithm (Figure 4.1) in that the Saturn system uses a dynamic control scheme to select the next most promising task. The D-algorithm separates the propagation of values into 2 serial steps: d-drive, followed by justification. In addition, the D-algorithm uses a static ordering criteria for choosing alternatives to achieve a task (d-drive or justification), and uses chronological backtracking in case of failure. On the other hand, the Saturn test generation system uses the deductive cost estimates to choose the cheapest alternative, and in case of failure backtracks to the choice that led to the failure, and not just the last choice.

The dynamic control architecture of Saturn has proved to be extremely useful in reducing the complexity of test generation (this will be discussed in greater detail in a later subsectio). This is especially true for sequential circuits where the dynamic control architecture can prevent the system from getting into infinite loops along cyclic paths.

4.3.2 Example

In this subsection we will present an example which illustrates the test generation algorithm outlined in the previous subsection for an adder device.

A picture of a design for the adder is given in Figure 4.3. At the top level the adder has two inputs and one output. Each input of the adder is 2 bits wide, and has two subports (one for each bit). Similarly, the output of the adder is 3 bits wide, and has three subports. This design partitions the adder into two full-adders, each of which is partitioned into

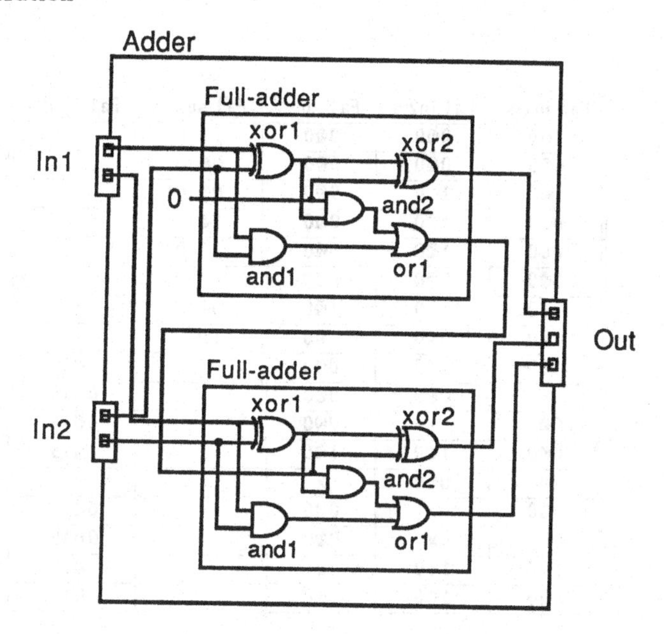

Figure 4.3: A design for an adder with 2 bit inputs.

an interconnection of *and*, *or*, and *xor* gates. The design specification includes the behavior for: the adder as a whole, the full-adders, the ten boolean gates, and the connections. In this design all gates have a delay of 5 time units, and the connections are 0 delay.

The user has specified that the inputs in1 and in2 of the adder are directly controllable, and the output out is directly observable. The user has not specified any test clichés, and the fault models are defined at the boolean level as stuck-at 1, and stuck-at 0 faults. In addition, assume that in order to minimize the number of tests generated the user has specified that all composite components are to be tested by testing their subparts.

At the gate level the test generator will generate tests that check if the inputs/output of the gates are stuck at a value of 1 (5 volts), or 0 (0 volts). For example, testing all stuck-at faults at the inputs and output of an *and* gate requires checking the output for the input combinations $< 1\ 1 >$, $< 0\ 1 >$, and $< 1\ 0 >$.

Since the adder is a composite module, it must be tested by testing its two full-adders. The full-adders themselves must be tested by testing their subparts which are the primitive boolean gates. For example, Figure 4.1 shows one test for one of the gate and2 of the first full-adder Fa1. When both inputs are 1 at time 0 the output must also be 1 at time 5.

Fa1-in1-s	Fa1-in2-s	Fa1-cin-s	Fa1-sum-s	Fa1-cout-s
1@0	0@0	0@0	1@10	
0@0	0@0	0@0	0@10	
1@0	1@0	0@0	0@10	
0@0	1@0	0@0	1@10	
0@0	1@0	0@0	1@10	
0@0	0@0	0@0	0@10	
0@0	1@0	1@0	0@10	
0@0	0@0	1@0	1@10	
1@0	1@0	0@0		1@15
x	0@0	0@0		0@15
0@0	x	0@0		0@15
0@0	1@0	1@0		1@15
0@0	0@0	x		0@15
0@0	x	0@0		0@15
0@0	x	0@0		0@15
1@0	1@0	x		1@15
0@0	1@0	1@0		1@15

Table 4.1: Uncompressed tests for a full-adder.

For this test, we must propagate the values at the inputs of and2 to the subports of the inputs of the full-adder, and propagate the value at the output of and2 to the subport of the carry output of the full-adder. In propagating the value at the output of the gate through the design we must control the values of other internal ports of the full-adder to ensure that this information is not masked, as would be the case, for example, if the second port of the *or* gate or1 is 1.

We can propagate these values through the design using the resolution residue inference procedure, which was described in the last chapter. The test for and2 propagated to the boundaries of the full-adder is: the first input is to be set to 1 at time -5, the second input is to be set to 0 at time -5, the carry input is to be set to 1 at time 0, and the carry output value 1 should be observed at time 10. To complete the test generation process for the full-adder we must achieve the remaining tests of and2, and all the tests for the remaining gates. These tests are given in Table 4.1.

The first three columns of the table correspond to the subports of the first three inputs of the full-adder, and the last two columns correspond to the subports of its two outputs. All variables in the table are assumed to be universally quantified. The variables in the input columns assert that the result is independent of the value for that input. The empty entries in the output columns correspond to unspecified values, i.e., we have not filled in the table for these entries.

The tests in the table have been *retimed*. For example, the test for and2 presented earlier corresponds to the sixth row from the bottom in the table. The time field of all values of a single test is shifted by the same amount so that the minimum time for any port value is 0. For this test, the time for the port values must be shifted by +5. Following this, the time for the first two inputs is 0, the time for the carry input is 5, and the time for the carry output is 15. Next, the first value for an input port is reset to time 0, e.g., the carry input is set to 1 at time 0, instead of time 5.

The quality of the tests generated by the system is inversely proportional to the length of the test vectors. An important step in the test generation process, thus, is to *compress* the tests in the previous table. Compressing a set of tests corresponds to combining individual tests whose inputs are compatible. Finding a minimal set of test vectors is equivalent to minimizing a collection of expressions, which is known to be NP-hard (possibly more difficult than NP-complete) for boolean expressions [22].

Instead of minimizing a collection of test vectors, the system employes a more computationally tractable, though non-minimal, compression algorithm (the cost is proportional to the square of the number of test vectors). The result of the algorithm is dependent upon the order of the input vectors. The algorithm compares each row in a table with the rows below it, starting from the top row. The actual comparison is done using the unification pattern matching process that was mentioned in the previous chapter. If the inputs of a row do not match any other row below it, it is left alone. If a row matches any row below it, it is deleted from the table and the variables in the other matching row are bound to the values that make the two rows equivalent. After a row is deleted, the matching process is continued with the row below it. The compression process stops when the bottom row in the table is reached.

The compressed tests for the previous table are given in Table 4.2. The original set of seventeen vectors has been reduced to six.

After compression, the tests must be *abstracted* to the next higher level in the hierarchy. In this case each port of the full-adder Fa1 has exactly one subport, and there is no transformation of the values across the port subport boundaries. The abstraction of these tests to the level

Fa1-in1-s	Fa1-in2-s	Fa1-cin-s	Fa1-sum-s	Fa1-cout-s
0@0	0@0	1@0	1@10	
0@0	1@0	1@0	0@10	1@15
0@0	1@0	0@0	1@10	
1@0	1@0	0@0	0@10	1@15
0@0	0@0	0@0	0@10	0@15
1@0	0@0	0@0	1@10	

Table 4.2: Compressed tests for a full-adder.

Fa1-in1	Fa1-in2	Fa1-cin	Fa1-sum	Fa1-cout
0@0	0@0	1@0	1@10	
0@0	1@0	1@0	0@10	1@15
0@0	1@0	0@0	1@10	
1@0	1@0	0@0	0@10	1@15
0@0	0@0	0@0	0@10	0@15
1@0	0@0	0@0	1@10	

Table 4.3: Compressed and abstracted tests for a full-adder.

of the full-adder results in the same values, but for the parent ports. This is not always the case. For example, for the adder the top-level behavior is defined in terms of integer values at the top-level ports, and the behavior of the subparts is defined in terms of boolean values for the subports of the adder. The abstracted tests for the full-adder Fa1 are given in Table 4.3.

The entire effort up to this point has been to define how to test the full-adder Fa1, given the definition of how to test its parts, and given their interconnection. Having done this work, we can share this definition for all components identical, or similar, to this full-adder, e.g., the other full-adder fa2. In order to share this definition the test vectors are *generalized.* The generalization, at present, is limited to apply to a class of components that only differ from the original in their delay. Other generalizations that we have not implemented include generalizing tests for components with similar behavior, e.g., where one behavior is a superset of the other, e.g., parameterizing tests for an adder based on the number of bits at each input. The generalized tests for the full-adder Fa1 are given in Table 4.4.

This test definition applies to arbitrary full-adders that have a delay

in1	in2	cin	sum	cout
0@0	0@0	1@0	1@d×2	
0@0	1@0	1@0	0@d×2	1@d×3
0@0	1@0	0@0	1@d×2	
1@0	1@0	0@0	0@d×2	1@d×3
0@0	0@0	0@0	0@d×2	0@d×3
1@0	0@0	0@0	1@d×2	

Table 4.4: Compressed, abstracted, and generalized tests for a full-adder.

of d time units for each gate. The port names in1 through cout must be replaced with the corresponding port names of the specific full-adder for which these definitions are to be used. This information is saved in a design library with the definition of the full-adder prototype, so that the definition can be shared across multiple designs.

The test generation task for the adder is not complete yet. We must achieve the abstracted tests, i.e., propagate them through the other full-adder fa2, and the connections, to the subports of the adder. The second full-adder fa2 must also be tested. The test vectors required to test it are the same as those for the first full-adder Fa1. In achieving the tests for the second full-adder, the values must be propagated through the first full-adder, and the connections, to the subports of the adder. In achieving the tests for the two full-adders we *never* propagate values through the substructure of the full-adders— the information is propagated using the top-level function definitions for the full-adders.

To complete the test generation for the adder, the tests at its subports must also be retimed, compressed, abstracted and generalized.

4.3.3 Control Strategies to Increase Efficiency

The previous subsection presented an example of test generation for a simple adder device, and outlined the control strategies that were used in increasing the efficiency. In this subsection we will examine the following control strategies in greater detail: conditional values, consistency checking, heuristics to guide search, and caching.

Conditional values provide a vocabulary for specifying if a value is conditional on the correct operation of the subpart being tested. By specifying conditional values we can improve the efficiency of test generation by discarding choices in the search space that correspond to useless tests. Consistency checking corresponds to checking if an alternative in the search space is consistent with the other subgoals of a test before examining it further. Heuristic search corresponds to using the deductive cost estimates to choose more promising alternatives in the search space first. Finally, caching corresponds to sharing test specifications and solutions to the subgoals of one test with other tests.

The remainder of this subsection will describe each of these control strategies in greater detail.

4.3.3.1 Conditional Values

Each local test for a subpart specifies the inputs and the expected output for these inputs. The output of the subpart is conditional on the presence of the fault being tested (i.e., the output is different if the fault is present). In achieving this test we have to propagate it to the directly controllable inputs/outputs. In order for a test to be useful the value propagated to a directly observable output of the device must be conditional on the fault being tested. For example, if the output is propagated through an *or* gate whose other input is 0, then the output is always 0 independent of the fault being tested.

Two ways of solving this problem are: to propagate a test to the directly controllable inputs/outputs and then check its utility, or to constrain the propagation of values to guarantee useful tests. The first alternative is not very attractive since propagating values is very expensive, and we may to repeat this process many times for each test. The second alternative is more attractive and is made possible by conditional values.

A conditional value is represented by the term [D <correct> <discrepancy>]. The "D" identifies the value as being conditional, where the condition is dependent on the functionality of the module being tested. If the module is functioning, the value is <correct>, and if it is not, the value is <discrepancy>. These discrepancy values are an extension of the d and $\bar{d}$ values of the D-algorithm to operate at arbitrary abstraction levels, and to take into account unknown discrepancy values (described later).

We can augment the behavior descriptions of components so that they describe how conditional values can be propagated through them. For example, for an *and* gate we can record the facts shown in Table 4.5.

The variables in this table are assumed to be universally quantified. The last rule corresponds to the case where the value to be made observable is propagated simultaneously from both inputs to the output.

(= Val(in1) [[D x y] t]) ∧ (= Val(in2) [true t]) ⇒ (= Val(out) [[D x y] +(t 5)])
(= Val(in2) [[D x y] t]) ∧ (= Val(in1) [true t]) ⇒ (= Val(out) [[D x y] +(t 5)])
(= Val(in1) [[D x y] t]) ∧ (= Val(in2) [[D x y] t]) ⇒ (= Val(out) [[D x y] +(t 5)])

Table 4.5: Conditional value propagation rules for an *and* gate.

These rules constrain the values of other arguments when propagating a test observable through a module. In propagating a test observable from an input of an *and* gate to its output, the other input must either be 1 or the same discrepancy value. It is possible to automatically deduce how to constrain the other arguments of a function in propagating conditional values through it.

Given some function f, and an argument i along which a test observable is to be propagated, we can compute the function f' which restricts the function f such that a conditional value is propagated to the output. Let $r_1, \ldots, r_n$ be the elements in the range of f which are under the image of more than one element in the domain of f. The inverse image of each r_j is some set S_j. We can sort the elements of S_j by the values of all the arguments, except the ith, to find the cases in which the ith argument is different and all other arguments are the same. For each such case, the domain of f is constrained such that each argument cannot have these values. The function f' is f with the conjunction of argument constraints for each r_j. It may be impossible to satisfy the conjunction of constraints simultaneously, in which case it is impossible to propagate a test observable from this argument of f.

For example, assume that we want to compute f' for the first argument of the *or* function. There is only one element in the range, 1, whose inverse image has more than one element. The set $S_{true} = \{< false\ true >, < true\ false >, < true\ true >\}$. Sorting this set by all arguments, except the first, gives:

$$< true > \rightarrow \{false, true\}$$
$$< false > \rightarrow \{true\}$$

That is, when the second argument is 1 the first argument can be 0 or 1, and when the second argument is 0 the first argument can only be 1. In this case, the second argument is constrained not to be 1.

The discrepancy element of a conditional value may be required in propagating multiple conditional values from the input of a module to its output. For example, if one argument of an adder is [D a b] we must ensure that the second argument is not [D b a] (since the result would always be +(a b) in this case). Knowing both the correct and discrepancy values may also be essential in propagating an observable from a single input of a module to its output. For example, if we are propagating the value 3 through a module which computes the *odd* predicate, then this can only be achieved if the discrepancy value is even. For boolean values, the discrepancy value is redundant, since it is the unique negation of the correct value. For more abstract objects, however, this is not the case. For example, for integer values, if we know that the correct value is 4, there is no unique value that is not 4.

In addition to augmenting the behavior of modules to propagate conditional values, the observables of a test must also be conditional values. At the boolean level it is easy to define the discrepancy value, given the correct value. However, this is not the case if tests are generated using the behavior description of a more abstract module where the domain size for port values is greater than two. In this situation there is a mismatch between the abstraction level of the behavior description and the level at which the fault models are meaningful. We can only say that the discrepancy value is some value in the domain other than the correct value. For example, suppose we are using the top-level behavior of an adder to define how to test it. If the inputs are 5 and 10 the correct output should be 15, and the discrepancy value can be any other. This test can be specified as:

(= Val(in1) [5 t]) ∧ (= Val(in2) [10 t]) ⇒ (= Val(out) [[D 15 ?] +(t 20)])

The conclusion of the test defines the correct value to be 15, and the discrepancy value to be ? (i.e., it is different from 15, but unknown). A similar problem arises when combining two tests (during the compression process) that have compatible inputs. Each test has a single correct value, and a single discrepancy value. The correct values for both tests are the same; however, the discrepancy values may be different since they are testing different faults. Unfortunately, the conditional value mechanism cannot record more than a single discrepancy value (since it applies to a single implicit condition vs. a set of conditions). We must record the discrepancy value using the same unknown value ?.

We can exploit the ability to propagate unknown discrepancy values to make the test generation process more efficient. For example, suppose we are propagating the conditional value [D true false] from the first input of an exclusive-or gate to its output. The exact value propagated to the output is dependent upon the value of the second input. If the second

```
(= Val(in1) [[D x y] t])∧(= Val(in2) [[N z] t])⇒(= Val(out) [[D ?1 ?2] +(t 5)])
(= Val(in2) [[D x y] t])∧(= Val(in1) [[N z] t])⇒(= Val(out) [[D ?1 ?2] +(t 5)])
(= Val(in1) [[D x y] t])∧(= Val(in2) [true t]) ⇒(= Val(out) [[D y x] +(t 5)])
(= Val(in2) [[D x y] t])∧(= Val(in1) [true t]) ⇒(= Val(out) [[D y x] +(t 5)])
(= Val(in1) [[D x y] t])∧(= Val(in2) [false t])⇒(= Val(out) [[D x y] +(t 5)])
(= Val(in2) [[D x y] t])∧(= Val(in1) [false t])⇒(= Val(out) [[D x y] +(t 5)])
```

Table 4.6: Propagating unknown conditional values.

input is 1 the output is [D false true], and if the second input is 0 the output is [D true false]. The output is always a conditional value as long as the second input is not a conditional value (the exclusive-or of [D x y] and [D x y] is 0, and the exclusive-or of [D x y] and [D y x] is 1). The actual conditional value at the output, however, is dependent on the value at the second input.

If we insist on knowing the exact conditional value at the output, we are forced to choose a value for the second input. We must control the second input of the gate to this value, and if it is inconsistent we are forced to backtrack. An alternative is to propagate an unknown conditional value at the output of the gate. In this case, both the correct and discrepancy values are unknown, but are guaranteed to be different. The conditional value propagation rules for an exclusive-or gate are given in Table 4.6.

All variables in the rules are assumed to be universally quantified. An unconditional value, whose value is unspecified, is represented as [N v]. The actual value will be bound to the variable v. The first rule above does not prematurely choose a particular value for the second input— it propagates an unknown conditional value at the output of the gate.

In achieving tests, we propagate the port values to be controlled to the inputs, and the port values to be observed to the outputs. All the input port values are completely defined (unconditional values), however, some outputs may be undefined conditional values. After compressing the tests, we can abstract the inputs of the test to the next level in the hierarchy, and simulate the high-level behavior with these abstracted inputs to compute the expected outputs.[1] We can use these expected outputs to fill in the correct value for the unknown conditionals at the

[1] We are simulating the high-level behavior, instead of simulating the collection of behaviors of the substructure, to improve the efficiency of computing the expected outputs.

output (the discrepancy value is still unknown).

It is not always possible to propagate conditional values through a module without knowing the actual values (e.g., the *odd* predicate example given earlier). Therefore, for completeness, we must also include the last four rules in the previous table. We can opportunistically try the top two rules first, and only try the other rules on failure.

4.3.3.2 Consistency Checking

In general, there will be more than one choice at each decision point in the search space. The goal of consistency checking is to prune inconsistent alternatives in the search space. In general, the problem of checking the consistency of an arbitrary collection of propositions is non-semi-decidable. Fortunately, the problem is decidable for a collection of propositions modeling a design of a finite-state machine (although it can be expensive). The actual propagation of values for consistency checking is done using the resolution inference procedure which was described in the previous chapter.

In choosing a subgoal sg_i at a node in the search space, consistency checking corresponds to seeing if it is possible to prove $\neg sg_i$, in which case this node is pruned. The utility of consistency checking is dependent upon selecting the appropriate amount effort in attempting to prove $\neg sg_i$. The drawback of too little effort is that inconsistencies are not caught early, and too much effort has the potential drawback that fruitless work may be done for consistent nodes. However, not all effort is lost for consistent nodes, since the facts derived during consistency checking are necessary conditions that must be true [20] for any solutions derived from the current node in the search space. These necessary conditions can be used to prune other inconsistent choices further down in the search space.

We can perform consistency checking either incrementally, or at the very end of test generation. The advantage of performing consistency checking incrementally at each decision point is that we can prune inconsistent alternatives immediately, thus saving the wasted effort of pursuing these paths further. In reasoning with sequential designs it is important to detect inconsistencies immediately, in order to detect identical recursive subgoals [54]. For combinational circuits, on the other hand, it is possible to perform the consistency checking at the very end, once a solution is found.

Performing consistency checking incrementally requires propagating the consequences of a choice through the design. These consequences can

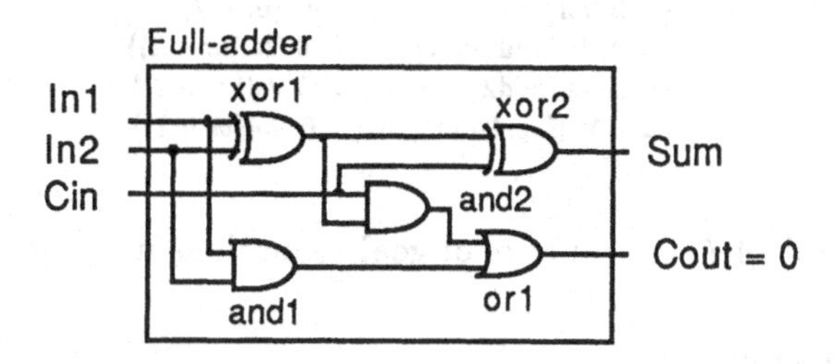

Figure 4.4: Controlling the output of a full-adder to 0.

propagate information forward and backward through the design. For example, if the output of an *and* gate is 0, and one input is 1, then the other input must be 0. Similarly, if one input of an *and* gate is 0, then the output must also be 0. We can improve the efficiency of consistency checking by terminating the propagation of information at the closest enclosing hierarchy boundary, since all information must pass through the ports at this boundary.

In general, it is computationaly inefficient to *prove* that a given choice is consistent. The Saturn system performs limited consistency checking by only propagating unit clauses. This is sufficient for detecting contradictions for atomic clauses, but not for disjunctive clauses. For example, it cannot detect that the four clauses $(a \vee b)$, $(\neg a \vee \neg b)$, $(\neg a \vee b)$, and $(a \vee \neg b)$ are mutually inconsistent.

Incremental consistency checking, thus, does not guarantee that a given choice is consistent. Some inconsistencies may be detected further from the point at which the erroneous choice was made. However, these inconsistencies will be detected by the time the values are propagated to the closest hierarchy boundary.

As an example illustrating consistency checking, consider controlling the output of the full-adder presented in Figure 4.4 to 0. The evolution of a partial solution is given in Table 4.7.

In this example we have chosen to control the output of and2 to false by controlling its second input to 0. At this point we have to decide how to control the output of the first exclusive-or gate to false. By examining the behavior of the first exclusive-or gate we find that one solution is to have both its inputs be 1. In checking if this choice is consistent we

(= 0 Val(cout))
(= 0 Val(or-out))
(= 0 Val(or-in1)) ∧ (= 0 Val(or-in2))
(= 0 Val(and2-out)) ∧ (= 0 Val(or-in2))
(= 0 Val(and2-in1)) ∧ (= 0 Val(or-in2))
(= 0 Val(xor1-out)) ∧ (= 0 Val(or-in2))

Table 4.7: A sequence of goal reductions using search.

conclude the following:

(= 1 Val(in1)) ∧ (= 1 Val(in2))⇒(= 1 Val(and1-in1)) ∧ (= 1 Val(and1-in2))
⇒(= 1 Val(or-in2))

This last fact is inconsistent, since it contradicts the second conjunct on the third line in the previous table. The other alternative of controlling both inputs of the first exclusive-or gate to 0, however, is consistent.

Since a given choice at a decision point may be inconsistent with the other facts in the design and the problem, we may have to revoke it and all its consequences. The Saturn system makes use of the *context* mechanism of MRS to partition facts whose justifications are based on assumptions. For nested assumptions we have a tree of contexts. Each new choice adds a child context to the current context, which is now the new current context. Only the facts in the current context and the facts in the contexts extending up the tree to the root are *active* at any given time. When an inconsistency is detected, the system decides the appropriate choice point to back up to. The facts in the contexts in the path from the current context to the context we are backtracking to are revoked. The current context now is the context we have backtracked to.

An advantage of performing consistency checking incrementally is that we can eliminate dependent subgoals in test generation. For example, suppose that a given task requires achieving the goal $a \wedge g$. In addition, assume that the subgoal g is dependent upon the subgoal a, i.e., $a \Rightarrow b \Rightarrow c \ldots \Rightarrow g$. In checking the consistency of the subtask a we will propagate its consequences, which include the other facts b through g. As these facts are propagated, they are added to the current context. After performing the consistency check for the subgoal a we can eliminate the subgoal g, since it is already true in the current context.

Incremental consistency checking is more expensive than consistency checking at the end, since it must be performed at every choice point in the search space, and at each such point it can require propagating information throughout the design. We can increase the efficiency of incremental consistency checking by only propagating values to the closest hierarchy boundary, instead of to the entire design (since the closest hierarchy boundary corresponds to the module for which we are generating tests). Performing consistency checking at the end, on the other hand, requires checking the solution once to ensure that a port is assigned only a single value at a given time. The selection of the appropriate strategy depends on the topology of the design for which tests are are being generated. Incremental consistency checking is appropriate for designs with a high degree of reconvergent fanout, since these are characterized by greater interaction between the subgoals of a task. Similarly, if a design has limited reconvergent fanout it may be more efficient to perform the consistency check at the end.

If an alternative at a choice point is inconsistent with the current assumptions, it is pruned from the search space. At this point we must decide where to backtrack to in the search tree. Since inconsistencies are not always detected immediately, chronological backtracking may be inefficient. One solution is to keep track of the justifications for each fact, and follow the justifications back to a choice point, then try another alternative for this choice. This corresponds to dependency-directed backtracking [18], which can have a dramatic impact on the efficiency of the problem solver.

4.3.3.3 Heuristics to Guide Search

Consistency checking cannot eliminate search, since there may be more than one consistent alternative at a choice point. The Saturn system uses heuristics to select the most promising alternative at a choice point. These heuristics are based on the estimated deductive cost for controlling/observing a port value, and the probability of being able to achieve this goal given the current problem constraints. These estimates do not take into account the actual values being controlled or observed. In addition, the system records with each port the fanout points and the directly controllable inputs which may need to be controlled to control/observe the value of this port.

The deductive cost estimates are integrated with the *agenda* mechanism of MRS to select the most promising alternative at a choice point. The agenda is a disjunction of tasks, where each task is a partial solution to achieve a test. Initially there is only one task, which specifies the test for the subpart being tested. The most promising task is removed from the agenda, and as it is reduced, other subtasks are added back on. The process of selecting a task from the agenda, and selecting a subtask of this task is defined below.

The cost of a task is equal to the sum of the costs of its subtasks. The cost of a subtask for controlling or observing a port value can be looked up directly in the knowledge base after the heuristic estimates have been calculated. For other subtasks the cost is assumed to be 1, e.g., for the subtask (= x and(true false)). The probability of being able to achieve a task depends on the interaction between its subtasks. These interactions are due to common fanout points in propagating information through the design. Each common fanout point represents a constraint (equivalence of values) that must be satisfied by all interacting subtasks. We can find the interactions by looking up the potential fanout points for each subtask (computed initially with the cost estimates), and checking for common points. The probability of achieving a task is equal to:

$$\prod_{i=1}^{commonfanouts} \frac{1}{D_i}$$

where D_i is the domain size of the values at the fanout point i.

We can use these cost and probability estimates to select the most promising task on the agenda. The tasks on the agenda are independent of each other, i.e., choosing a particular alternative for one task does not restrict, or influence, the other tasks. Let $C(T_i)$ be the cost of executing task T_i, let $P(T_i)$ be the probability of succeeding in achieving task T_i, and $C(T_i;T_j)$ be the cost of executing task T_i followed by executing task T_j. Then:

$$C(T_i;T_j) = C(T_i) + (1 - P(T_i)) \times C(T_j)$$
$$C(T_i;T_j) < C(T_j;T_i) \iff \frac{P(T_i)}{C(T_i)} > \frac{P(T_j)}{C(T_j)}$$
$$P(T_i) \approx P(T_j) \Rightarrow C(T_i;T_j) < C(T_j;T_i) \iff C(T_i) < C(T_j)$$

These equations define a total ordering for the tasks on the agenda, and they can be used to select the best task to work on next. Having selected the best task, we must still select the most promising conjunct of this task to work on next.[2] This corresponds to ordering conjuncts [53]

[2] In the CNF representation, this corresponds to selecting a literal of the task to match with other clauses in the knowledge base.

in problem solving, which can have a dramatic impact on the cost. The subtasks of a task are assumed to be highly interdependent, i.e., choosing a particular solution for one can constrain the possible solutions to the other subtasks (due to common fanout nodes). Consequently, we will be using the heuristic which selects the conjunct with the fewest estimated solutions first. That is:

$$C(sg_i; sg_j) < C(sg_j; sg_i) \iff Numsol(sg_i) < Numsol(sg_j)$$

The number of solutions to a subtask is equal to the number of combinations for the directly controllable inputs that achieve it. Each subtask is only dependent upon some of the directly controllable inputs. Some fraction of all the combinations of these dependent inputs actually achieve the subtask. The number of solutions is equal to this number multiplied by all possible combinations for the other inputs that the subtask is independent of. As a heuristic, we will assume that the number of solutions is inversely proportional to the number of directly controllable inputs that a subtask is dependent upon. This information can be looked up directly in the knowledge base after the cost heuristics have been computed. We will, therefore, select the subtask which is dependent on the most inputs.

Algorithm for Computing Cost Estimates

The remainder of this subsection will outline the algorithm for computing the cost heuristics, and will show the results of this algorithm for a full-adder. In addition, we will present experimental results that illustrate the advantage of using these cost heuristics to guide search.

The algorithm computes the worst case estimates for the cost of controlling and observing the value of a port. It first computes the cost of controlling the ports by propagating information from the directly controllable inputs to the directly observable outputs. In the second pass, it computes the cost of observing ports by propagating information from the directly observable outputs to the directly controllable inputs.

As we compute the cost of controlling ports, we will also keep track of the directly controllable inputs and the fanout points which may need to be controlled in controlling the value of this port.

The cost of controlling a directly controllable input is 1 by default. If the user prefers controlling certain inputs over others, these number can be changed. For example, if certain inputs are harder to control than others, they can be assigned a higher cost. In addition, in controlling this port, the dependent directly controllable inputs are the port itself, and the dependent fanout nodes are the port itself, if it is a fanout node.

The cost of controlling a subport of an input port is one more than the cost of controlling the parent port. In controlling this subport to some value, the dependent directly controllable inputs and fanout nodes are the same as those for the parent port.

Similarly, the cost of controlling the end port of a connection is one more than the cost of controlling the starting port of the connection. In controlling the end port of the connection, the dependent directly controllable inputs and fanouts are the same as those for the starting port. However, at a fanout, the dependent fanout nodes for controlling the end ports also include the starting port of the connection.

The cost of controlling an output of a module is one more than the sum of the costs for controlling its inputs, and the dependent directly controllable inputs and fanouts are the union of these for all the input ports. If a module has multiple outputs, the dependent fanout nodes for each output also include the inputs of the module.

The algorithm, as it has been described thus far, is adequate for computing the control costs for designs without feedback paths. However, for designs with feedback paths we cannot compute the cost of controlling the output of a module, since the cost of controlling the feedback inputs is undefined. Using a finite-state machine model, we assume that controlling the output to some value requires cycling through half the states of the machine, on the average. Each cycle of the finite state machine requires controlling its inputs and the feedback paths. We know the cost of controlling the inputs, however, the cost of controling the feedback paths must be computed. This is achieved by propagating a tagged value at the outputs of the module with 0 cost. As these values are propagated through the design, the cost is updated. Eventually these values will be propagated back to the feedback inputs of the module, and we can compute the cost of controlling the feedback paths. By using a tree of tagged value types we can break the feedback path of arbitrarily interconnected finite state machines.

The following equation defines the control costs for arbitrary modules:

$$\#states_i \times \left(1 + \sum_{k=1}^{\#inputs} ControlCost(input_k)\right)$$

that is, the cost of controlling the output is one plus the sum of controlling the inputs of the module (which includes the feedback paths), multiplied by the number of internal states of the module. For combinational circuits, the factor outside the parenthesis is one. Having computed the cost estimates for controlling the ports (and the dependent directly controllable inputs and fanouts in controlling the ports), we can compute the cost estimates for observing the ports. As we compute the cost of observing ports, we will also keep track of the directly controllable inputs and the fanout points that may need to be controlled in observing the value of this port.

The cost of observing a directly observable output is 1 by default. If the user prefers observing certain outputs over others, these numbers can be changed. For example, if certain outputs are harder to observe than others, they can be assigned a higher cost. In addition, in observing a value at this port there are no dependent directly controllable inputs or fanouts.

The cost of observing a subport of an output port is one more than the cost of observing the parent port. In observing a value at this subport the dependent directly controllable inputs and fanout nodes are the same as those for the parent port.

Similarly, the cost of observing the starting port of a connection is one more than the cost of observing the ending port of the connection. In observing the starting port of a connection, the dependent directly controllable inputs and fanouts are the same as those for the ending port. However, at a fanout, the cost of observing the starting port is the minimum of the cost of observing each of the endpoints of the fanout, plus one. In addition, the dependent directly controllable inputs and fanouts for observing the value at the fanout port are the same as those for the endpoint of the connection with the minimum cost.

The cost of observing an input of a module is equal to one plus the sum of the costs of controlling the other inputs, plus the cost of observing the output with the minimum cost. We have ignored the internal states of the module since any conditional value at an output is acceptable. The dependent directly controllable inputs and fanouts for observing an input of a module are the union of these for all the other inputs and the selected output.

As an example, Figure 4.5 illustrates the cost of controlling the ports of a full-adder. Similarly, the cost of observing the ports of the full-adder is given in Figure 4.6.

These estimates assign a lower cost to more abstract design descrip-

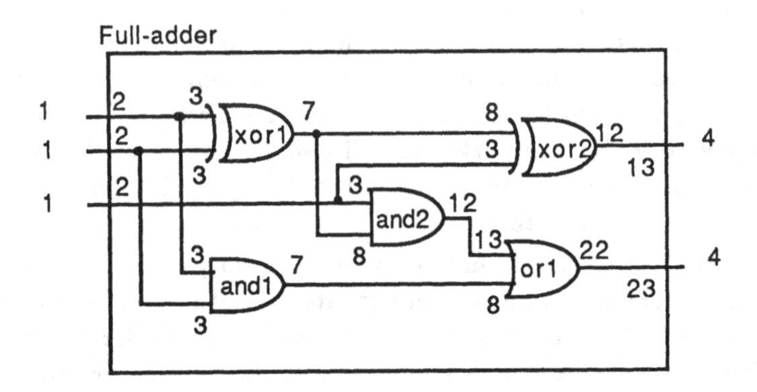

Figure 4.5: Costs for controlling the ports of a full-adder.

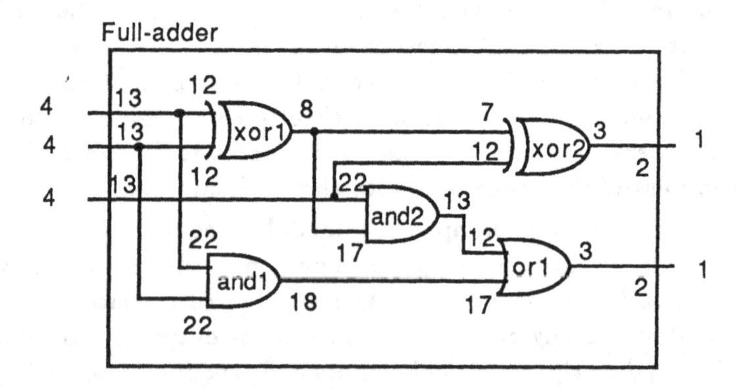

Figure 4.6: Costs for observing the ports of a full-adder.

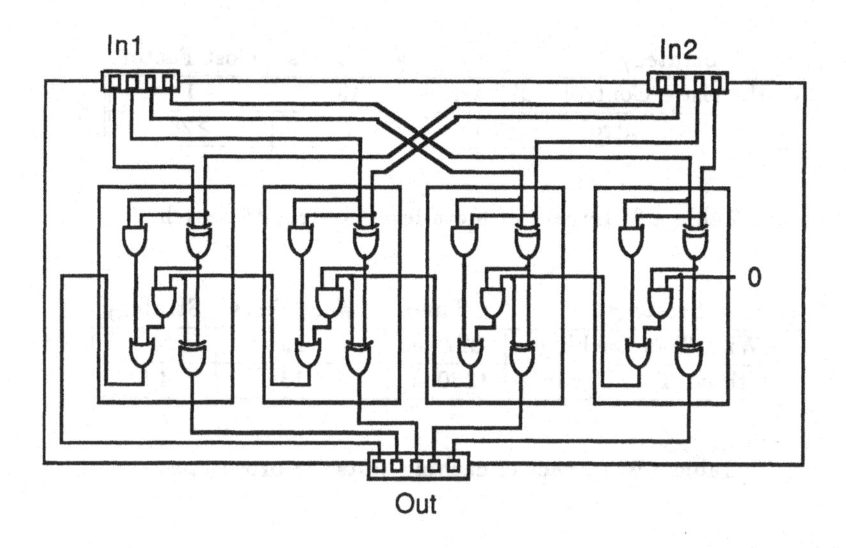

Figure 4.7: A picture of a design for an adder with 4 bit inputs.

tions. For example, the cost of controlling the sum output is cheaper than the cost of controlling the subport of the sum output. In controlling the parent port we use the top-level behavior of the full-adder. However, in controlling the subport we must use the behavior of the submodules and connections of the full-adder. A similar situation arises in observing a value at an input port of the full-adder. The cost of observing the parent port is cheaper than that of observing the value of its subport.

To examine the utility of guiding search with these heuristics we generated tests for an adder with, and without, the run-time control for selecting the best task, and the best subtask of this task. Each input of the adder is four bits wide, and the design for the adder is defined in terms of four full-adders. Each full-adder itself is defined in terms of a collection of boolean gates. A picture of this adder is given in Figure 4.7, and the summary of the test generation results is given in Table 4.8.

Static control corresponds to choosing the tasks based on the order in which the behavior rules are defined in the knowledge base, and choosing the first conjunct in each rule. For this design there are 118 components, 25 modules, and 73 tests which have been compressed to 13.

Strategy	Time	# of Tests	Cost Factor
Heuristic Control	272 sec.	13	1
Static Control	≥2 hr.	?	≥26

Table 4.8: Impact of meta-level control on search.

Strategy	Time	# of Tests	Strategy
With Test Caching	272 sec.	13	1
Without Test Caching	1230 sec.	13	4.5

Table 4.9: Impact of caching tests for prototypes.

4.3.3.4 Caching

Caching is used to share the results of one task across different tasks. In the Saturn test generation system caching is used to share the definition of how to test one component with other similar components, and to share the solutions to subtasks for one test with other tests.

The test generation process is divided into two steps. In the first step we must decide how to test a module, and in the second step we must achieve these tests (propagate them to the closest hierarchy boundary).

Caching tests for composite components corresponds to sharing results for the first step of test generation. For example, for the design pictured in Figure 4.7, after generating the definition of how to test the first full-adder (by testing its parts), we can share this definition with the other three instances of the full-adders in the design.

In general, without test caching, the number of tests to generate is proportional to the number of modules in the design. For a hierarchically formulated design with l levels and approximately n modules per level, the number of modules to generate tests for is approximately n^l. However, with test caching, the number of tests to generate is proportional to the number of distinct module types in the design. For a large design with a small set of module types the difference can be very dramatic.

Table 4.9 summarizes the results for generating tests with, and without, caching tests:

Caching solutions to subtasks corresponds to sharing the results for the second step in test generation, i.e., in achieving tests. Test generation

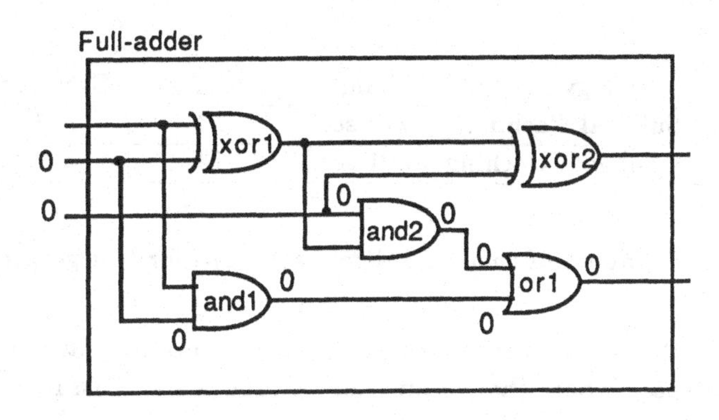

Figure 4.8: A test for the *or* gate.

involves testing all the possible faults for a device. These tasks share many subtasks in common. For example, a test for the *or* gate of a full-adder is given in Figure 4.8:

In this test both inputs must be controlled to 0. The first input can be set to 0 by controlling the carry input to 0. We can cache this solution with the other two subtasks for controlling the output and the first input of and2 to 0. One of the subtasks in testing and2 requires controlling its first input to 0. Instead of generating a solution to this goal again, we can look it up directly if it is cached. Similarly, the solutions to the task of making port values observable can be shared across different tests.

As solutions to subgoals are cached the system dynamically updates the cost for achieving these goals. For example, after finding the first solution S_i to the subgoal i the cost of achieving the subgoal i is changed to 1, since a solution for it can be looked up directly without requiring inference. Consequently, the meta-level control chooses cached solutions first, before exploring alternate ways of reducing a subgoal. The cost of the subgoals of i will be greater than 1, unless there is exactly one subgoal of i for which a solution is cached.

In addition to caching solutions to subgoals we can also cache the fact that there are no solutions to a conjunction of subgoals. In a state-space representation each node in the search space includes the list of subgoals to achieve in order to solve the original goal. In addition, with each node we record the set of assumptions that have been made thus far in reducing the original goal. If a node in the search space is inconsistent, then the conjunction of the subgoals and assumables for this node are

Strategy	Time	# of Tests	Cost Factor
With Subgoal Caching	272 sec.	13	1
Without Subgoal Caching	591 sec.	13	2.2

Table 4.10: Impact of caching subgoal solutions for test generation.

inconsistent.[3] We can cache this result to enable pruning other nodes in the search space where the conjunction of subgoals and assumables is a superset of those for this node.

Caching can be disadvantageous, however, if there are no solutions for a test. For example, it is impossible to control both inputs of the *or* gate of a full-adder to 1. If a node in the search space has the conjunction of subgoals $x \wedge y \wedge z$, and the subgoal x has cached solutions, then each cached solution duplicates the search space for the remaining subgoals $y \wedge z$. Cached solutions for the other subgoals have a similar multiplicative affect on the size of the search space, which increases approximately by the average number of cached solutions for each subgoal raised to the power of the number of subgoals with cached solutions. The most dramatic impact of this is when there is no solution to the original goal, in which case the entire expanded search space must be explored. Fortunately, most tests are achievable and we do not run into this problem.

Table 4.10 compares the experimental results for generating tests with, and without, subgoal caching for the four bit adder presented earlier:

4.3.4 Empirical Evaluation

In this section we will present empirical results that illustrate the utility of reasoning with reformulated designs. We will first show the utility of refining a design, and next the utility of abstracting a design.

In generating tests for a device we must exercise all behavior that might be modified by the possible faults in the device. If the fault models are unknown, or if the fault models do not apply at the granularity of the design description, we must test the behavior of the design completely (all input/state combinations) to ensure that all possible faults

[3] Actually, a subset of the subgoals and assumables may be inconsistent. However, finding this minimal set would require additional effort.

Device	# Tests Flat	# Tests Refined	Time Flat	Time Refined
full-adder	16	9	.84 sec.	33 sec.
$adder_2$	16	9	.6 sec.	82 sec.
$adder_3$	64	12	1.8 sec.	121 sec.
$adder_4$	256	13	6.6 sec.	190 sec.
$adder_8$	65536[a]	13	>1800 sec.[a]	772 sec.

[a](ran out of swap space on a 3670)

Table 4.11: Utility of refining designs.

are tested. For designs with a large number of inputs this can get quite expensive. However, if the fault models are known, and the design description exists at the level of the possible faults, we can test a device by testing each fault, thereby generating fewer tests.

Assume that the possible failures for a device are stuck-at faults for the ports at the boolean level (and we are assuming single faults). That is, the ports of a boolean gate can either be stuck-at 0 or stuck-at 1. Table 4.11 compares the results of generating tests for different devices using a flat high-level design formulation (a black box) versus using a formulation where the design description is refined to the level of boolean gates. The examples are for a full-adder and adders with varying number of bits at each input.

The tests for both design formulations completely test all possible failures in the device for the given fault model. Although the number of tests generated using the more refined design formulation is far smaller, this is not at the expense of completeness. The reduction in the number of tests for the refined design is made possible by compressing the individual tests at the hierarchy boundary (tests with compatible inputs are combined). The number of test vectors required to test a device with the flat high-level design formulation grows exponentially with the number of components. However, both the number of tests and the growth rate is much smaller for the refined design.

The last two columns show the time required to generate the tests using the flat high-level and refined design formulations. For the flat design formulation the actual time is smaller for small designs, but grows exponentially. The slight reduction in going from the full-adder to the adder is due to the speedup in using the *add* instruction of the machine

Formulation	Time	# Tests	Cost Factor
Hierarchical	272 sec.	13	1
Flat Gate Level	962 sec.	16	3.5

Table 4.12: Advantage of abstracting a design description.

on which the system is running, instead of using a collection of rules to enumerate all input/output combinations. For this implementation, the crossover point occurs for adders with 7 bit inputs. For complex devices, the flat design formulation provides us with poorer solutions (longer test vectors) at a greater cost.

As an example illustrating the utility of abstracting a design consider the adder presented earlier in Figure 4.7. The original flat design formulation describes the adder at the level of gates only. The reformulated design imposes a hierarchical structure on top of the original design by first abstracting gates into full-adders, and next abstracting the 4 full-adders to define the adder.

The advantage of reasoning with abstracted design formulations is that we can propagate information over a collection of modules in one step, without requiring any backtracking, etc. The test generator shifts to the next higher design formulation at every abstraction hierarchy. The more abstract the level of the design, the greater the savings in propagating information at this level.

Table 4.12 presents the results of generating tests for these two design formulations. By using the reformulated design the cost of test generation is reduced by a factor of 3.5 (for complex devices, such factors are multiplied for each additional level in the hierarchy). Interestingly, in addition to the savings in time, fewer test vectors are generated with the abstracted design. This is due to the imperfections in the test compression algorithm, which compromises minimality for efficiency (the minimization problem is NP-hard). The repeated application of test compression at each hierarchy boundary, with fewer tests, generates a shorter test vector length than a single application with a large set of vectors.

An example of test generation for a more realistic design is given in Appendix B for the printer adapter card of an IBM PC. The interesting aspects of this device are that it includes: components with states (register, latches), feedback paths, bi-directional signals, and tri-state signals (busses). The original design description provided in the IBM technical manual for this card was manually reformulated to capture a higher level description. By exploiting this high level design description the Saturn test generation system took approximately 30 minutes to test this card.

Chapter 5

Conclusion

This chapter will present a summary of the key ideas described in this thesis, suggest directions for further research in reformulating designs, and describe the implementation state of the Saturn test generation system.

5.1 Summary of Key Ideas

In this thesis we have examined the utility of reasoning with high-level design formulations of devices. We have restricted ourselves to devices whose structure can be partitioned into a collection of modules with well defined behavior, that interact with each other along specific channels of communication, e.g., digital circuits.

Creating and maintaining such systems requires reasoning about the device for a collection of tasks, including simulation, verification, test generation, and diagnosis. In order to reason about the device we must formulate a design for the device, encode a description of this design in a representation language, and use an inference procedure to reason with this description.

A design is a specification of a device at a collection of abstraction levels. A specification at each abstraction level defines: a partitioning of the device into a collection of modules, the communication channels between these modules, and the behavior for the modules and the communication channels.

In order to be useful, the design specifications at the different abstraction levels must be a correct specification of the device. The structural correctness between two levels of a design, or between a design level and the device, requires defining a mapping between the modules at one level and the corresponding set of modules at the lower abstraction level, or the set of physical components in the device. Similarly, behavioral correctness requires defining a homomorphism between the composition of the behaviors of the modules at the lower level and the behavior of the parent of these components, or between the behavior of the device and behavior of the corresponding components in the design.

In order to manage the complexity of reasoning about a device we should capture it morphology in its design description. The morphology of the device corresponds to its subparts, and the relationships that exist between these parts. The process of designing a device is usually partitioned into a collection of refinement operations between different abstraction levels. The morphology of the device reflects the organization of the design at these abstraction levels.

The formulation of a design can have a dramatic impact on the cost of the reasoning about the device. By capturing the morphology of the device we can reduce the depth of the search space, and also reduce the branching factor of the nodes in the search space. As a consequence, the size of the search space can be reduced exponentially (e.g., for test generation). By reformulating designs we can increase the level of complexity of the devices that we can handle for a given task.

We have empirically validated the utility of reasoning with high-level design formulations for test generation by demonstrating that both the time and the quality of solutions can be improved by reasoning with a design that captures the morphology of the device. Similar advantages have been obtained for other reasoning tasks, e.g., simulation [52] and diagnosis [23].

By using device independent representation methods, we can use the same design description of a device for a collection of tasks, e.g., simulation, diagnosis, and test generation. In addition, we can exploit the generality of the representation methods to encode partial descriptions of devices. By using general inference methods it is possible to use the same inference procedure to reason about different parts of the the device at different abstraction levels, and to use the same inference procedure for different tasks. This generality has a cost overhead associated with it. For small designs this generality will lead to an inefficiency. However, for large designs it can mean the difference between being able to solve a problem or not.

It is essential to have control over the reasoning procedure in order to guide the system to choose the more fruitful alternatives in the search space in a situation dependent manner. It is possible to increase the efficiency of the reasoning procedure further by incorporating domain specific control strategies, e.g., conditional values, consistency checking, heuristics to guide search, and caching for test generation.

Although the ideal design formulation can be dependent upon both the task and the device, a useful formulation to consider initially is the one created in the design process. If only a low-level design formulation is available, it is appropriate to manually and/or automatically reformulate the design to capture higher-level design descriptions.

The main contributions of this thesis are: the analysis of the different types of reformulation operations, and the advantages of performing these reformulations; the definition of a correctness criteria for reformulating designs; the creation of a task independent language for describing digital designs (Corona); the creation of the Saturn test generation system which is unique in that it exploits design descriptions at arbitrary abstraction levels to generate tests efficiently; and the empirical validation of the utility of reasoning with high level design descriptions for a real world design.

5.2 Further Work

In this thesis we have examined the utility of reasoning with high-level design formulations. Although we have described strategies for automatically reformulating designs we have not mechanized any of these. Having defined the various types of reformulation operations, a logical direction for further research would be to attempt to automate these reformulations.

It would be easier to develop such a theory of reformulations for a restricted domain, e.g., digital circuits. Ideally such a theory would define the reformulation operations to perform given the initial design formulation, and the specification of the types of reasoning involved in the task. For example, for test generation, it is useful to define the rules for propagating conditional values from the inputs of modules to their output, in addition to abstracting the structure and behavior of the design.

In this thesis we have only examined the utility of different design formulations for analytic reasoning tasks, e.g., simulation, test generation, and diagnosis. Another fruitful area of research would be to extend the analysis of reformulation for synthetic tasks, e.g., discovering the important types of knowledge, and their organization, for automatically refining designs.

5.3 Implementation State

The Saturn test generation system has been implemented in Maclisp [46] running on a PDP-20, and Symbolics 3600 series machines. An integral part of the reasoning and representation methods of the system are built on top of those provided by the MRS [49] knowledge representation system. The inputs to the system are the specifications of the structure of the device, the behavior for the subparts, rules for propagating conditional values, a specification of the possible faults, and optionally user specified test cliches which include the conditions under which they should be applied.

We have generated tests for the device D74 using the design formulation presented in Figure 5.1. This formulation of the device has approximately 650 objects, seven levels of hierarchy, and data objects ranging from bits to integers. We have also generated tests for the Printer Adapter card of the IBM PC (see Appendix A and B), which represents a simple, though real design. The interesting aspects of this board are that it has feedback paths, bi-directional signals, and tri-state busses. In addition, the behavior for the modules in this design are described in terms of list encodings of bit vectors.

The system is fully implemented except for the constraint solver. Such a constraint satisfaction system will have to simultaneously solve a set of equations, where each equation is a composition of the behavior of the modules in the design, e.g., arbitrary boolean expressions.

Appendix A

Printer Adapter Card

This appendix presents the design description for the IBM PC Printer Adapter card. The design is described in the language CORONA, which is a prefixed form of Predicate Calculus. The syntax of the language is different from the examples presented in the thesis— its definition can be found in the CORONA language manual [51].

The Printer Adapter card has two interfaces: one to the main processor bus of the personal computer, and the other to the parallel port of the printer. The card accepts commands from the host computer and translates them into appropriate control and data signals at the printer bus. The printer transmits its status (out of paper, character printed, etc.) information back to the card, which interrupts the host when appropriate.

Figure A.1 shows a picture of the design of the IBM PC Printer Adapter card as defined in the IBM technical reference manual. This design is a flat formulation of the device at the boolean level, in terms of boolean gates and TTL integrated circuits.

The tests for this device were generated from an alternate design formulation, which is pictured in Figure A.2. In this design we have partitioned the card into the following 8 modules: command decoder, data transceiver, data buffer, control buffer, control latch, data latch, control driver, and the local bus. In this design formulation the behavior of the components is described at a more abstract level than the original boolean level. The signal values at the ports of the modules are list data structures encoding the boolean values of the lower abstraction level.

Every module at this abstract level includes: its behavior specification, the specifications for propagating conditional values from an input to an output, and the specification of how this module should be tested at the abstract level. The fault model for this device is stuck at faults at the boolean level. Since the device is specified at a level more abstract than the fault model, the default is to test the modules by testing their behavior exhaustively. Unfortunately, this can get very expensive, specially in testing all the bit combinations for the data paths. The part of the design that specifies how to test a module corresponds to user supplied test cliche's which are used in place of exhaustively testing the behavior of a module.

The important properties of this design are: the behavior is specified at an abstract level, the behavior of the modules includes temporal information (delays were obtained from the TTL catalog), the design includes bi-directional and tri-state signals (for the local bus, etc.), and the design includes sequential behavior (due to the feedback paths, and the state in the latches and registers).

The design defines the top level module corresponding to the printer adapter card (named pa), the ports of the card, the submodules, and the interconnection for these submodules. In addition, the design includes a specification of which ports can be directly controlled and which can be directly observed. In achieving a test, the values are propagated through the design to these ports. Finally, the design includes a specification of the behavior, conditional value propagation rules, and test information for each of the 8 prototypes that are the submodules of the printer adapter card.

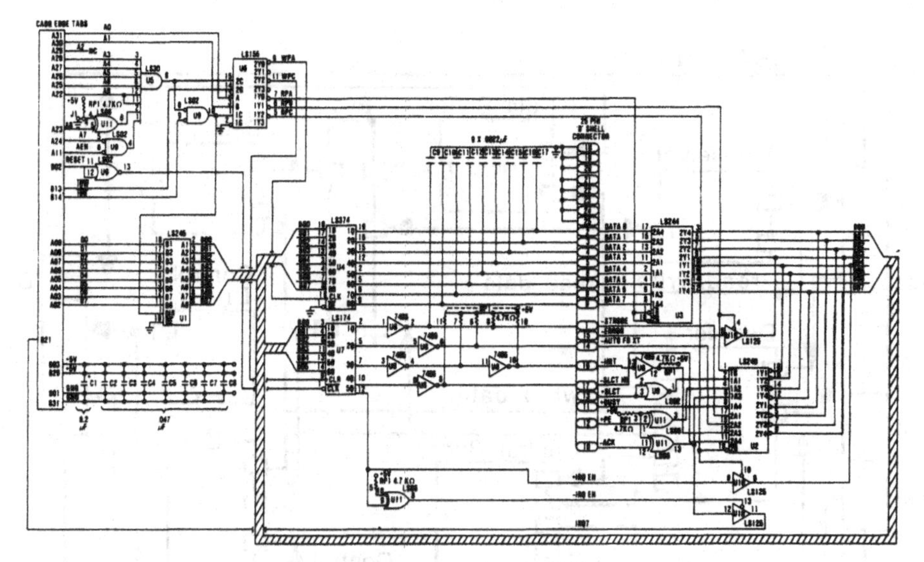

Figure A.1: The IBM specification for the Printer Adapter card.

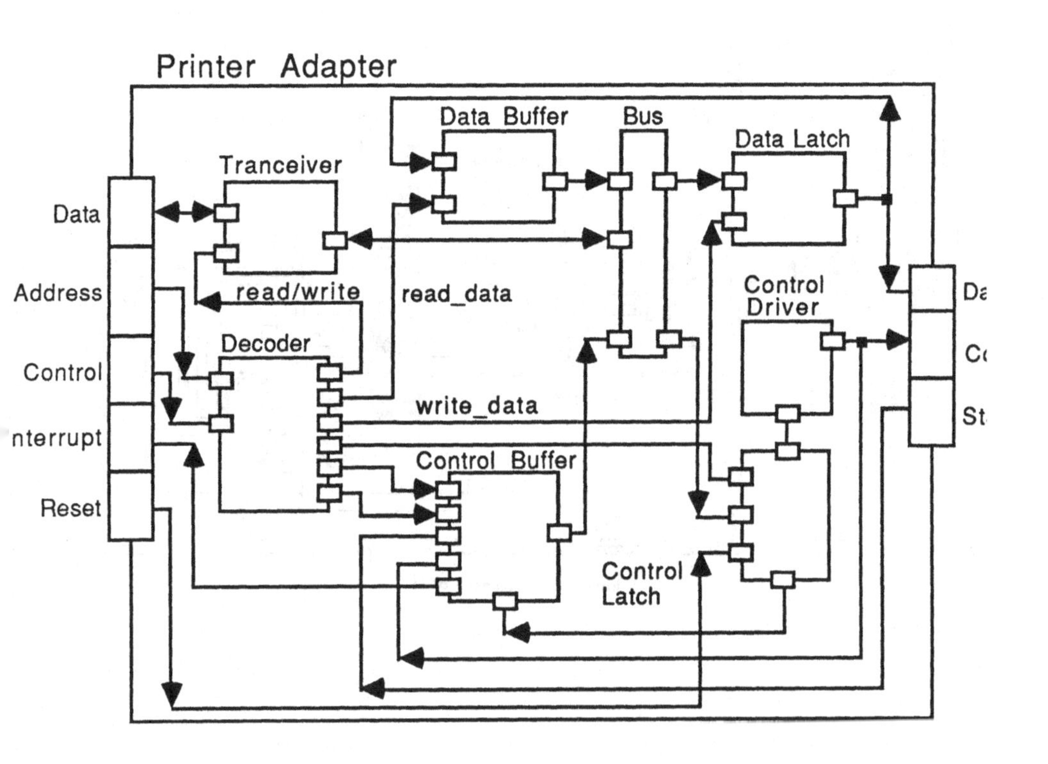

Figure A.2: An abstract formulation of the Printer Adapter card.

```
;;;;;;;;;;;;;;;;;;;;;;;;;;;;;;;;;;;;;;;;;;;;;;;;;;;;;;;;;;;;;;;;;;;;;;;;;;;;
;;;       Corona definition for the IBM PC printer Adapter card.         ;;;
;;;       The top level module is named "pa".                            ;;;
;;;;;;;;;;;;;;;;;;;;;;;;;;;;;;;;;;;;;;;;;;;;;;;;;;;;;;;;;;;;;;;;;;;;;;;;;;;;

; Structural description.

; First the port definitions.

; First the motherboard port.
(port bus pa) ; i/o interface port.
(dsize (port bus pa) 67108864)

(subport bus-data bus pa)
(input bus-data pa)
(output bus-data pa)
(dsize (port bus-data pa) 256)

(subport bus-address bus pa)
(input bus-address pa)
(dsize (port bus-address pa) 1024) ; 10 bits of address

(subport bus-control bus pa)
(input bus-control pa)
(dsize (port bus-control pa) 64)

(subport bus-ir bus pa)
(output bus-ir pa)
(dsize (port bus-ir pa) 2)

(subport bus-reset bus pa)
(input bus-reset pa)
(dsize (port bus-reset pa) 2)

; Next, the printer port.
(port pbus pa)
(dsize (port pbus pa) 131072)

(subport pbus-data pbus pa)
(output pbus-data pa)
(dsize (port pbus-data pa) 256)

(subport pbus-control pbus pa)
(output pbus-control pa)
(dsize (port pbus-control pa) 16)

(subport pbus-status pbus pa)
(input pbus-status pa)
(dsize (port pbus-status pa) 32)

; Module definitions.
(subpart pa-decoder pa)   ; command decoder
(module pa-decoder)
(type (module pa-decoder) cdecoder)

(subpart pa-rectran pa)   ; transceiver
(module pa-rectran)
(type (module pa-rectran) transceiver)

(subpart pa-busdbuff pa)  ; data bus buffer
(module pa-busdbuff)
(type (module pa-busdbuff) dbuffer)
```

```
(subpart pa-buscbuff pa)  ; control bus buffer
(module pa-buscbuff)
(type (module pa-buscbuff) cbuffer)

(subpart pa-clatch pa)    ; control latch
(module pa-clatch)
(type (module pa-clatch) clatch)

(subpart pa-dlatch pa)    ; data latch
(module pa-dlatch)
(type (module pa-dlatch) dlatch)

(subpart pa-cdriver pa)   ; control driver
(module pa-cdriver)
(type (module pa-cdriver) driver)

(subpart pa-localbus pa)  ; local bus
(module pa-localbus)
(type (module pa-localbus) (lbus 5))

; Specify the top-level connections.

; Connections for the command decoder.
(conn (port bus-address pa) (port address pa-decoder))
(conn (port bus-control pa) (port control pa-decoder))

(conn (port dir pa-decoder) (port dir pa-rectran))
(conn (port rdata pa-decoder) (port enable pa-busdbuff))
(conn (port wdata pa-decoder) (port clock pa-dlatch))
(conn (port wcontrol pa-decoder) (port clock pa-clatch))
(conn (port rstatus pa-decoder) (port enable1 pa-buscbuff))
(conn (port rcontrol pa-decoder) (port enable2 pa-buscbuff))

; Connections for the transceiver.
(conn (port data1 pa-rectran) (port bus-data pa))
(conn (port bus-data pa) (port data1 pa-rectran))
(conn (port data2 pa-rectran) (port 3 pa-localbus))
(conn (port 3 pa-localbus) (port data2 pa-rectran))

; Connections for the Bus Data Buffer.
(conn (port data2 pa-busdbuff) (port 1 pa-localbus))
(conn (port data2 pa-dlatch) (port data1 pa-busdbuff))

; Connections for the Data Latch.
(conn (port 2 pa-localbus) (port data1 pa-dlatch))
(conn (port data2 pa-dlatch) (port pbus-data pa))

; Connections for the Control Bus Buffers.
(conn (port control pa-cdriver) (port control pa-buscbuff))
(conn (port pbus-status pa) (port status pa-buscbuff))
(conn (port irq pa-clatch) (port irq pa-buscbuff))
(conn (port irq7 pa-buscbuff) (port bus-ir pa))
(conn (port data pa-buscbuff) (port 4 pa-localbus))

; Connections for the Control Latch.
(conn (port bus-reset pa) (port clear pa-clatch))
(conn (port data2 pa-clatch) (port data pa-cdriver))
(conn (port 5 pa-localbus) (port data1 pa-clatch))

; Connections for the Control Drivers.
(conn (port control pa-cdriver) (port pbus-control pa))

; Define setable and observable ports for test generation.
```

```
(setable (port bus-address pa) (n $n $n))
(setable (port bus-control pa) (n $n $n))
(setable (port bus-reset pa) (n $n $n))
(setable (port bus-data pa) (n $n $n))
(setable (port pbus-status pa) (n $n $n))

(observable (port bus-data pa) (d $n $d))
(observable (port bus-ir pa) (d $n $d))
(observable (port pbus-control pa) (d $n $d))
(observable (port pbus-data pa) (d $n $d))
```

```
;;;;;;;;;;;;;;;;;;;;;;;;;;;;;;;;;;;;;;;;;;;;;;;;;;;;;;;;;;;;;;;;;;;;;;;;;;;;;;
;;;        Corona definitions for the command decoder prototype          ;;;
;;;;;;;;;;;;;;;;;;;;;;;;;;;;;;;;;;;;;;;;;;;;;;;;;;;;;;;;;;;;;;;;;;;;;;;;;;;;;;

; Top-level module definitions.

; Define the structure for the command decoder.
(if (type (module $mod) cdecoder)
    (and (port address $mod) (input address $mod)
         (dsize (port address $mod) 524288)
         (port control $mod) (input control $mod)
         (dsize (port control $mod) 64)
         (port dir $mod) (output dir $mod) (dsize (port dir $mod) 2)
         (port rdata $mod) (output rdata $mod)
         (dsize (port rdata $mod) 2) (dsize (port wdata $mod) 2)
         (port wdata $mod) (output wdata $mod)
         (port wcontrol $mod) (output wcontrol $mod)
         (dsize (port wcontrol $mod) 2)
         (port rstatus $mod) (output rstatus $mod)
         (dsize (port rstatus $mod) 2)
         (port rcontrol $mod) (output rcontrol $mod)
         (dsize (port rcontrol $mod) 2)))

; Default behavior preferences:
(if (type (module $mod) cdecoder) (bpref (module $mod) $x normal rule))

; Behavior for Command Decoder.
(if (type (module $mod) cdecoder)
    (brules (module $mod) $x normal

; define when the board is not addressed.
(if (or (true (val (port address $mod)
                   (n (0 $b $c $d $e $f $g $h $i $j)
                      (0 $b $c $d $e $f $g $h $i $j))) $t)
        (true (val (port address $mod)
                   (n ($a 0 $c $d $e $f $g $h $i $j)
                      ($a 0 $c $d $e $f $g $h $i $j))) $t)
        (true (val (port address $mod)
                   (n ($a $b 1 $d $e $f $g $h $i $j)
                      ($a $b 1 $d $e $f $g $h $i $j))) $t)
        (true (val (port address $mod)
                   (n ($a $b $c 0 $e $f $g $h $i $j)
                      ($a $b $c 0 $e $f $g $h $i $j))) $t)
        (true (val (port address $mod)
                   (n ($a $b $c $d 0 $f $g $h $i $j)
                      ($a $b $c $d 0 $f $g $h $i $j))) $t)
        (true (val (port address $mod)
                   (n ($a $b $c $d $e 0 $g $h $i $j)
                      ($a $b $c $d $e 0 $g $h $i $j))) $t)
        (true (val (port address $mod)
                   (n ($a $b $c $d $e $f 0 $h $i $j)
                      ($a $b $c $d $e $f 0 $h $i $j))) $t))
    (unaddressed $mod $t))

; Define what it means to have a delay of 100 time units for this module.
; patterns have been compiled for this module.
(if (and (delay-100 (val (port $p $mod) (n 1 1)) $t)
         (= $t1 (+ 100 $t)))
    (true (val (port $p $mod) (n 1 1)) $t1))

; define the "rdata" output.
(if (and (true (val (port address $mod)
                    (n (1 1 0 1 1 1 1 $h 0 0)
```

```
                    (1 1 0 1 1 1 1 $h 0 0))) $t)
          (true (val (port control $mod)
                     (n ($a1 $b1 0 $d1 $e1 $f1)
                        ($a1 $b1 0 $d1 $e1 $f1))) $t)
          (= $t1 (+ $t 100)))
     (true (val (port rdata $mod) (n 0 0)) $t1))

(if (or (unaddressed $mod $t)
        (true (val (port control $mod)
                   (n ($a1 $b1 1 $d1 $e1 $f1)
                      ($a1 $b1 1 $d1 $e1 $f1))) $t)
        (true (val (port address $mod)
                   (n ($a $b $c $d $e $f $g $h $i 1)
                      ($a $b $c $d $e $f $g $h $i 1))) $t)
        (true (val (port address $mod)
                   (n ($a $b $c $d $e $f $g $h 1 $j)
                      ($a $b $c $d $e $f $g $h 1 $j))) $t))
     (delay-100 (val (port rdata $mod) (n 1 1)) $t))

; define the "rstatus" output.
(if (and (true (val (port address $mod)
                    (n (1 1 0 1 1 1 1 $h 0 1)
                       (1 1 0 1 1 1 1 $h 0 1))) $t)
         (true (val (port control $mod)
                    (n ($a1 $b1 0 $d1 $e1 $f1)
                       ($a1 $b1 0 $d1 $e1 $f1))) $t)
         (= $t1 (+ $t 100)))
     (true (val (port rstatus $mod) (n 0 0)) $t1))

(if (or (unaddressed $mod $t)
        (true (val (port control $mod)
                   (n ($a1 $b1 1 $d1 $e1 $f1)
                      ($a1 $b1 1 $d1 $e1 $f1))) $t)
        (true (val (port address $mod)
                   (n ($a $b $c $d $e $f $g $h $i 0)
                      ($a $b $c $d $e $f $g $h $i 0))) $t)
        (true (val (port address $mod)
                   (n ($a $b $c $d $e $f $g $h 1 $j)
                      ($a $b $c $d $e $f $g $h 1 $j))) $t))
     (delay-100 (val (port rstatus $mod) (n 1 1)) $t))


; define the "rcontrol" output.
(if (and (true (val (port address $mod)
                    (n (1 1 0 1 1 1 1 $h 1 0)
                       (1 1 0 1 1 1 1 $h 1 0))) $t)
         (true (val (port control $mod)
                    (n ($a1 $b1 0 $d1 $e1 $f1)
                       ($a1 $b1 0 $d1 $e1 $f1))) $t)
         (= $t1 (+ $t 100)))
     (true (val (port rcontrol $mod) (n 0 0)) $t1))

(if (or (unaddressed $mod $t)
        (true (val (port control $mod)
                   (n ($a1 $b1 1 $d1 $e1 $f1)
                      ($a1 $b1 1 $d1 $e1 $f1))) $t)
        (true (val (port address $mod)
                   (n ($a $b $c $d $e $f $g $h $i 1)
                      ($a $b $c $d $e $f $g $h $i 1))) $t)
        (true (val (port address $mod)
                   (n ($a $b $c $d $e $f $g $h 0 $j)
                      ($a $b $c $d $e $f $g $h 0 $j))) $t))
     (delay-100 (val (port rcontrol $mod) (n 1 1)) $t))
```

```
; define the "dir" output.
(if (and (true (val (port address $mod)
                    (n (1 1 0 1 1 1 1 $h $i $j)
                       (1 1 0 1 1 1 1 $h $i $j))) $t)
         (true (val (port control $mod)
                    (n ($a1 $b1 0 $d1 $e1 $f1)
                       ($a1 $b1 0 $d1 $e1 $f1))) $t)
         (= $t1 (+ $t 100)))
    (true (val (port dir $mod) (n 0 0)) $t1))

(if (or (unaddressed $mod $t)
        (true (val (port control $mod)
                   (n ($a1 $b1 1 $d1 $e1 $f1)
                      ($a1 $b1 1 $d1 $e1 $f1))) $t))
    (delay-100 (val (port dir $mod) (n 1 1)) $t))

; define the "wdata" output.
(if (and (true (val (port address $mod)
                    (n (1 1 0 1 1 1 1 $h 0 0)
                       (1 1 0 1 1 1 1 $h 0 0))) $t)
         (true (val (port control $mod)
                    (n ($a1 $b1 $c1 0 $e1 $f1)
                       ($a1 $b1 $c1 0 $e1 $f1))) $t)
         (= $t1 (+ $t 100)))
    (true (val (port wdata $mod) (n 0 0)) $t1))

(if (or (unaddressed $mod $t)
        (true (val (port control $mod)
                   (n ($a1 $b1 $c1 1 $e1 $f1)
                      ($a1 $b1 $c1 1 $e1 $f1))) $t)
        (true (val (port address $mod)
                   (n ($a $b $c $d $e $f $g $h $i 1)
                      ($a $b $c $d $e $f $g $h $i 1))) $t)
        (true (val (port address $mod)
                   (n ($a $b $c $d $e $f $g $h 1 $j)
                      ($a $b $c $d $e $f $g $h 1 $j))) $t))
    (delay-100 (val (port wdata $mod) (n 1 1)) $t))

; define the "wcontrol" output.
(if (and (true (val (port address $mod)
                    (n (1 1 0 1 1 1 1 $h 1 0)
                       (1 1 0 1 1 1 1 $h 1 0))) $t)
         (true (val (port control $mod)
                    (n ($a1 $b1 $c1 0 $e1 $f1)
                       ($a1 $b1 $c1 0 $e1 $f1))) $t)
         (= $t1 (+ $t 100)))
    (true (val (port wcontrol $mod) (n 0 0)) $t1))

(if (or (unaddressed $mod $t)
        (true (val (port control $mod)
                   (n ($a1 $b1 $c1 1 $e1 $f1)
                      ($a1 $b1 $c1 1 $e1 $f1))) $t)
        (true (val (port address $mod)
                   (n ($a $b $c $d $e $f $g $h $i 1)
                      ($a $b $c $d $e $f $g $h $i 1))) $t)
        (true (val (port address $mod)
                   (n ($a $b $c $d $e $f $g $h 0 $j)
                      ($a $b $c $d $e $f $g $h 0 $j))) $t))
    (delay-100 (val (port wcontrol $mod) (n 1 1)) $t))
))
```

```
;;;;;;;;;;;;;;;;;;;;;;;;;;;;;;;;;;;;;;;;;;;;;;;;;;;;;;;;;;;;;;;;;;;;;;;;;;;;;;;;
;;;                Define how to test the command decoder                   ;;;
;;;;;;;;;;;;;;;;;;;;;;;;;;;;;;;;;;;;;;;;;;;;;;;;;;;;;;;;;;;;;;;;;;;;;;;;;;;;;;;;

(if (type (module $mod) cdecoder)
    (trules (module $mod) behavior ; if testing via top-level.

; tests for the "rdata" output.
(if (and (true (val (port address $mod) (n (1 1 0 1 1 1 1 $h 0 0)
                                           (1 1 0 1 1 1 1 $h 0 0))) 0)
         (true (val (port control $mod)
                    (n ($a1 $b1 0 $d1 $e1 $f1)
                       ($a1 $b1 0 $d1 $e1 $f1))) 0))
    (true (val (port rdata $mod) (d 0 1)) 100))

(if (true (val (port control $mod) (n ($a1 $b1 1 $d1 $e1 $f1)
                                       ($a1 $b1 1 $d1 $e1 $f1))) 0)
    (true (val (port rdata $mod) (d 1 0)) 100))

(if (true (val (port address $mod)
               (n ($a $b $c $d $e $f $g $h $i 1)
                  ($a $b $c $d $e $f $g $h $i 1))) 0)
    (true (val (port rdata $mod) (d 1 0)) 100))


; tests for the "rstatus" output.
(if (and (true (val (port address $mod)
                    (n (1 1 0 1 1 1 1 $h 0 1)
                       (1 1 0 1 1 1 1 $h 0 1))) 0)
         (true (val (port control $mod)
                    (n ($a1 $b1 0 $d1 $e1 $f1)
                       ($a1 $b1 0 $d1 $e1 $f1))) 0))
    (true (val (port rstatus $mod) (d 0 1)) 100))

(if (true (val (port control $mod) (n ($a1 $b1 1 $d1 $e1 $f1)
                                       ($a1 $b1 1 $d1 $e1 $f1))) 0)
    (true (val (port rstatus $mod) (d 1 0)) 100))

(if (true (val (port address $mod)
               (n ($a $b $c $d $e $f $g $h 1 $j)
                  ($a $b $c $d $e $f $g $h 1 $j))) 0)
    (true (val (port rstatus $mod) (d 1 0)) 100))

; tests for the "rcontrol" output.
(if (and (true (val (port address $mod) (n (1 1 0 1 1 1 1 $h 1 0)
                                           (1 1 0 1 1 1 1 $h 1 0))) 0)
         (true (val (port control $mod) (n ($a1 $b1 0 $d1 $e1 $f1)
                                           ($a1 $b1 0 $d1 $e1 $f1))) 0))
    (true (val (port rcontrol $mod) (d 0 1)) 100))

(if (true (val (port control $mod) (n ($a1 $b1 1 $d1 $e1 $f1)
                                       ($a1 $b1 1 $d1 $e1 $f1))) 0)
    (true (val (port rcontrol $mod) (d 1 0)) 100))

(if (true (val (port address $mod)
               (n ($a $b $c $d $e $f $g $h $i 1)
                  ($a $b $c $d $e $f $g $h $i 1))) 0)
    (true (val (port rcontrol $mod) (d 1 0)) 100))

; tests for the "dir" output.
(if (and (true (val (port address $mod)
                    (n (1 1 0 1 1 1 1 $h $i $j)
                       (1 1 0 1 1 1 1 $h $i $j))) 0)
```

```
        (true (val (port control $mod)
                   (n ($a1 $b1 0 $d1 $e1 $f1)
                      ($a1 $b1 0 $d1 $e1 $f1))) 0))
    (true (val (port dir $mod) (d 0 1)) 100))

(if (true (val (port control $mod) (n ($a1 $b1 1 $d1 $e1 $f1)
                                       ($a1 $b1 1 $d1 $e1 $f1))) 0)
    (true (val (port dir $mod) (d 1 0)) 100))

(if (true (val (port address $mod) (n (0 1 0 1 1 1 1 $h $i $j)
                                       (0 1 0 1 1 1 1 $h $i $j))) 0)
    (true (val (port dir $mod) (d 1 0)) 100))

; tests for the "wdata" output.
(if (and (true (val (port address $mod)
                   (n (1 1 0 1 1 1 1 $h 0 0)
                      (1 1 0 1 1 1 1 $h 0 0))) 0)
         (true (val (port control $mod)
                   (n ($a1 $b1 $c1 0 $e1 $f1)
                      ($a1 $b1 $c1 0 $e1 $f1))) 0))
    (true (val (port wdata $mod) (d 0 1)) 100))

(if (true (val (port control $mod)
               (n ($a1 $b1 $c1 1 $e1 $f1)
                  ($a1 $b1 $c1 1 $e1 $f1))) 0)
    (true (val (port wdata $mod) (d 1 0)) 100))

(if (true (val (port address $mod)
               (n ($a $b $c $d $e $f $g $h 1 $j)
                  ($a $b $c $d $e $f $g $h 1 $j))) 0)
    (true (val (port wdata $mod) (d 1 0)) 100))

; tests for the "wcontrol" output.
(if (and (true (val (port address $mod)
                   (n (1 1 0 1 1 1 1 $h 1 0)
                      (1 1 0 1 1 1 1 $h 1 0))) 0)
         (true (val (port control $mod)
                   (n ($a1 $b1 $c1 0 $e1 $f1)
                      ($a1 $b1 $c1 0 $e1 $f1))) 0))
    (true (val (port wcontrol $mod) (d 0 1)) 100))

(if (true (val (port control $mod)
               (n ($a1 $b1 $c1 1 $e1 $f1)
                  ($a1 $b1 $c1 1 $e1 $f1))) 0)
    (true (val (port wcontrol $mod) (d 1 0)) 100))

(if (true (val (port address $mod)
               (n ($a $b $c $d $e $f $g $h $i 1)
                  ($a $b $c $d $e $f $g $h $i 1))) 0)
    (true (val (port wcontrol $mod) (d 1 0)) 100))

))
```

```
;;;;;;;;;;;;;;;;;;;;;;;;;;;;;;;;;;;;;;;;;;;;;;;;;;;;;;;;;;;;;;;;;;;;;;;;
;;;        Corona definitions for the transceiver prototype          ;;;
;;;;;;;;;;;;;;;;;;;;;;;;;;;;;;;;;;;;;;;;;;;;;;;;;;;;;;;;;;;;;;;;;;;;;;;;

; Structure description for transceivers.
(if (type (module $mod) transceiver)
    (and (port data1 $mod) (input data1 $mod) (output data1 $mod)
         (dsize (port data1 $mod) 256)
         (port data2 $mod) (input data2 $mod) (output data2 $mod)
         (dsize (port data2 $mod) 256)
         (port dir $mod) (input dir $mod) (dsize (port dir $mod) 2)))

; Default preferences.
(if (type (module $mod) transceiver)
        (bpref (module $mod) $x normal rule))

; Behavior for transceivers.
(if (type (module $mod) transceiver)
    (brules (module $mod) $x normal

            (if (and (true (val (port dir $mod) (n 1 1)) $t)
                     (true (val (port data1 $mod) $v) $t)
                     (= $t1 (+ $t 30)))
                (true (val (port data2 $mod) $v) $t1))

            (if (and (true (val (port dir $mod) (n 0 0)) $t)
                     (true (val (port data2 $mod) $v) $t)
                     (= $t1 (+ $t 30)))
                (true (val (port data1 $mod) $v) $t1))

            (if (and (true (val (port dir $mod) (d 1 0)) $t)
                     (true (val (port data1 $mod)
                                (n (1 0 1 0 1 0 1 0)
                                   (1 0 1 0 1 0 1 0))) $t)
                     (= $t1 (+ $t 30)))
                (true (val (port data2 $mod)
                           (d (1 0 1 0 1 0 1 0)
                              (1 1 1 1 1 1 1 1))) $t1))

            (if (and (true (val (port dir $mod) (d 0 1)) $t)
                     (true (val (port data2 $mod)
                                (n (1 0 1 0 1 0 1 0)
                                   (1 0 1 0 1 0 1 0))) $t)
                     (= $t1 (+ $t 30)))
                (true (val (port data1 $mod)
                           (d (1 0 1 0 1 0 1 0)
                              (1 1 1 1 1 1 1 1))) $t1))
))
```

```
;;;;;;;;;;;;;;;;;;;;;;;;;;;;;;;;;;;;;;;;;;;;;;;;;;;;;;;;;;;;;;;;;;;;;;;;
;;;                Define how to test the transceiver                ;;;
;;;;;;;;;;;;;;;;;;;;;;;;;;;;;;;;;;;;;;;;;;;;;;;;;;;;;;;;;;;;;;;;;;;;;;;;

(if (type (module $mod) transceiver)
    (trules (module $mod) behavior ; if testing via top-level.

; test going from data1 to data2.
(if (and (true (val (port dir $mod) (n 1 1)) 0)
         (true (val (port data1 $mod)
                    (n (1 0 1 0 1 0 1 0) (1 0 1 0 1 0 1 0))) 0))
    (true (val (port data2 $mod) (d (1 0 1 0 1 0 1 0)
                                    (1 1 1 1 1 1 1 1))) 30))

(if (and (true (val (port dir $mod) (n 1 1)) 0)
         (true (val (port data1 $mod)
                    (n (0 1 0 1 0 1 0 1) (0 1 0 1 0 1 0 1))) 0))
    (true (val (port data2 $mod) (d (0 1 0 1 0 1 0 1)
                                    (1 1 1 1 1 1 1 1))) 30))

; test going from data2 to data1.
(if (and (true (val (port dir $mod) (n 0 0)) 0)
         (true (val (port data2 $mod)
                    (n (1 0 1 0 1 0 1 0) (1 0 1 0 1 0 1 0))) 0))
    (true (val (port data1 $mod) (d (1 0 1 0 1 0 1 0)
                                    (1 1 1 1 1 1 1 1))) 30))

(if (and (true (val (port dir $mod) (n 0 0)) 0)
         (true (val (port data2 $mod)
                    (n (0 1 0 1 0 1 0 1) (0 1 0 1 0 1 0 1))) 0))
    (true (val (port data1 $mod) (d (0 1 0 1 0 1 0 1)
                                    (1 1 1 1 1 1 1 1))) 30))

))
```

```
;;;;;;;;;;;;;;;;;;;;;;;;;;;;;;;;;;;;;;;;;;;;;;;;;;;;;;;;;;;;;;;;;;;;;;;;;;;;
;;;       Corona definitions for the data buffer prototype             ;;;
;;;;;;;;;;;;;;;;;;;;;;;;;;;;;;;;;;;;;;;;;;;;;;;;;;;;;;;;;;;;;;;;;;;;;;;;;;;;

; Structure description for data bus buffers.
(if (type (module $mod) dbuffer)
    (and (port data1 $mod) (input data1 $mod)
         (dsize (port data1 $mod) 256) (dsize (port data2 $mod) 256)
         (port data2 $mod) (output data2 $mod)
         (port enable $mod) (input enable $mod)
         (dsize (port enable $mod) 2)))

; Default preferences.
(if (type (module $mod) dbuffer) (bpref (module $mod) $x normal rule))

; Behavior for data bus buffers.
(if (type (module $mod) dbuffer)
    (brules (module $mod) $x normal

; if enabled, the output follows the input.
            (if (and (true (val (port enable $mod) (n 0 0)) $t)
                     (true (val (port data1 $mod) $v) $t)
                     (= $t1 (+ $t 30)))
               (true (val (port data2 $mod) $v) $t1))

            (if (and (true (val (port enable $mod) (d 0 1)) $t)
                     (true (val (port data1 $mod)
                                (n (1 0 1 0 1 0 1 0)
                                   (1 0 1 0 1 0 1 0))) $t)
                     (= $t1 (+ $t 30)))
                (true (val (port data2 $mod)
                           (d (1 0 1 0 1 0 1 0)
                              (1 1 1 1 1 1 1 1))) $t))

            (if (and (true (val (port enable $mod) (d 1 0)) $t)
                     (true (val (port data1 $mod)
                                (n (1 0 1 0 1 0 1 0)
                                   (1 0 1 0 1 0 1 0))) $t)
                     (= $t1 (+ $t 30)))
                (true (val (port data2 $mod)
                           (d (1 1 1 1 1 1 1 1)
                              (1 0 1 0 1 0 1 0))) $t))
))
```

```
;;;;;;;;;;;;;;;;;;;;;;;;;;;;;;;;;;;;;;;;;;;;;;;;;;;;;;;;;;;;;;;;;;;;;;;;;;;;;;;
;;;               Define how to test the data buffer prototype             ;;;
;;;;;;;;;;;;;;;;;;;;;;;;;;;;;;;;;;;;;;;;;;;;;;;;;;;;;;;;;;;;;;;;;;;;;;;;;;;;;;;

(if (type (module $mod) dbuffer)
    (trules (module $mod) behavior ; if testing via top-level.


; test the output for 2 different inputs, when enabled.
(if (and (true (val (port enable $mod) (n 0 0)) 0)
         (true (val (port data1 $mod)
                    (n (1 0 1 0 1 0 1 0) (1 0 1 0 1 0 1 0))) 0))
    (true (val (port data2 $mod) (d (1 0 1 0 1 0 1 0)
                                    (1 1 1 1 1 1 1 1))) 30))

(if (and (true (val (port enable $mod) (n 0 0)) 0)
         (true (val (port data1 $mod)
                    (n (0 1 0 1 0 1 0 1) (0 1 0 1 0 1 0 1))) 0))
    (true (val (port data2 $mod) (d (0 1 0 1 0 1 0 1)
                                    (1 1 1 1 1 1 1 1))) 30))
))
```

```
;;;;;;;;;;;;;;;;;;;;;;;;;;;;;;;;;;;;;;;;;;;;;;;;;;;;;;;;;;;;;;;;;;;;;;;;;
;;;     Corona definitions for the control buffer prototype          ;;;
;;;;;;;;;;;;;;;;;;;;;;;;;;;;;;;;;;;;;;;;;;;;;;;;;;;;;;;;;;;;;;;;;;;;;;;;;

; Define the structure for the control buffer.
(if (type (module $mod) cbuffer)
    (and (port enable1 $mod) (input enable1 $mod)
         (dsize (port enable1 $mod) 2) (dsize (port enable2 $mod) 2)
         (port enable2 $mod) (input enable2 $mod)
         (port status $mod) (input status $mod)
         (dsize (port status $mod) 32) (dsize (port control $mod) 16)
         (port control $mod)(input control $mod)
         (port irq $mod) (input irq $mod) (dsize (port irq $mod) 2)
         (port irq7 $mod) (output irq7 $mod) (dsize (port irq7 $mod) 2)
         (port data $mod) (output data $mod)
         (dsize (port data $mod) 256)))

; Default preferences.
(if (type (module $mod) cbuffer) (bpref (module $mod) $x normal rule))

; Define the behavior for the control buffer.
(if (type (module $mod) cbuffer)
    (brules (module $mod) $x normal

; define the behavior for the irq7 port.
         (if (and (true (val (port irq $mod) (n 1 1)) $t)
                  (true (val (port status $mod)
                             (n ($a $b $c $d $e)
                                ($a $b $c $d $e))) $t)
                  (bnot $b $v1)
                  (= $t1 (+ $t 60)))
             (true (val (port irq7 $mod) (n $v1 $v1)) $t1))

         (if (and (true (val (port irq $mod) (n 0 0)) $t)
                  (= $t1 (+ $t 60)))
             (true (val (port irq7 $mod) (n 0 0)) $t1))

; define behavior when enable1 is true (reading status).
         (if (and (true (val (port enable1 $mod) (n 0 0)) $t)
                  (true (val (port status $mod)
                             (n ($a $b $c $d $e)
                                ($a $b $c $d $e))) $t)
                  (bnot $a $v)
                  (= $t1 (+ $t 60)))
             (true (val (port data $mod)
                        (n ($v $b $c $d $e 1 1 1)
                           ($v $b $c $d $e 1 1 1))) $t1))

; define behavior when enable2 is true (reading control).
         (if (and (true (val (port enable2 $mod) (n 0 0)) $t)
                  (true (val (port control $mod)
                             (n ($a $b $c $d)
                                ($a $b $c $d))) $t)
                  (bnot $d $di)
                  (bnot $c $ci)
                  (bnot $a $ai)
                  (true (val (port irq $mod) (n $i $i)) $t)
                  (= $t1 (+ $t 60)))
             (true (val (port data $mod)
                        (n (1 1 1 $i $ai $b $ci $di)
                           (1 1 1 $i $ai $b $ci $di)))$t1))

; define the "d" rules.
```

```
            (if (and (true (val (port irq $mod) (d $a $b)) $t)
                     (true (val (port status $mod)
                                (n (0 0 0 0 0)
                                   (0 0 0 0 0))) $t)
                     (= $t1 (+ $t 60)))
                (true (val (port irq7 $mod) (d $a $b)) $t1))

            (if (and (true (val (port enable1 $mod) (d 0 1)) $t)
                     (true (val (port status $mod)
                                (n (1 0 0 0 0)
                                   (1 0 0 0 0))) $t)
                     (= $t1 (+$t 60)))
                (true (val (port data $mod)
                           (d (0 0 0 0 0 1 1 1)
                              (1 1 1 1 1 1 1 1))) $t1))

            (if (and (true (val (port enable1 $mod) (d 1 0)) $t)
                     (true (val (port status $mod)
                                (n (1 0 0 0 0) (1 0 0 0 0)))
                           $t)
                     (= $t1 (+$t 60)))
                (true (val (port data $mod)
                           (d (1 1 1 1 1 1 1 1)
                              (0 0 0 0 0 1 1 1))) $t1))

            (if (and (true (val (port enable2 $mod) (d 0 1)) $t)
                     (true (val (port control $mod)
                                (n (1 0 1 1) (1 0 1 1))) $t)
                     (true (val (port irq $mod) (n 0 0)) $t)
                     (= $t1 (+$t 60)))
                (true (val (port data $mod)
                           (d (1 1 1 0 0 0 0 0)
                              (1 1 1 1 1 1 1 1))) $t1))

            (if (and (true (val (port enable2 $mod) (d 1 0)) $t)
                     (true (val (port control $mod)
                                (n (1 0 1 1) (1 0 1 1))) $t)
                     (true (val (port irq $mod) (n 0 0)) $t)
                     (= $t1 (+$t 60)))
                (true (val (port data $mod)
                           (d (1 1 1 1 1 1 1 1)
                              (1 1 1 0 0 0 0 0))) $t1))

))
```

```
;;;;;;;;;;;;;;;;;;;;;;;;;;;;;;;;;;;;;;;;;;;;;;;;;;;;;;;;;;;;;;;;;;;;;;;;;
;;;              Define how to test the control buffer               ;;;
;;;;;;;;;;;;;;;;;;;;;;;;;;;;;;;;;;;;;;;;;;;;;;;;;;;;;;;;;;;;;;;;;;;;;;;;;

(if (type (module $mod) cbuffer)
    (trules (module $mod) behavior ; if testing via top-level.

; test the irq7 port.
            (if (and (true (val (port irq $mod) (n 1 1)) 0)
                     (true (val (port status $mod)
                                (n ($a 0 $c $d $e)
                                   ($a 0 $c $d $e))) 0))
                (true (val (port irq7 $mod) (d 1 0)) 60))

            (if (true (val (port irq $mod) (n 0 0)) 0)
                (true (val (port irq7 $mod) (d 0 1)) 60))

; test reading status with enable1.
            (if (and (true (val (port enable1 $mod) (n 0 0)) 0)
                     (true (val (port status $mod)
                                (n (1 1 0 1 0)
                                   (1 1 0 1 0))) 0))
                (true (val (port data $mod)
                           (d (0 1 0 1 0 1 1 1)
                              (1 1 1 1 1 1 1 1))) 60))

; test reading control with enable2.
            (if (and (true (val (port enable2 $mod) (n 0 0)) 0)
                     (true (val (port control $mod)
                                (n (0 0 0 1) (0 0 0 1))) 0)
                     (true (val (port irq $mod) (n 0 0)) 0))
                (true (val (port data $mod)
                           (d (1 1 1 0 1 0 1 0)
                              (1 1 1 1 1 1 1 1))) 60))
))
```

```
;;;;;;;;;;;;;;;;;;;;;;;;;;;;;;;;;;;;;;;;;;;;;;;;;;;;;;;;;;;;;;;;;;;;;;;;;;;;;
;;;        Corona definitions for the control latch prototype            ;;;
;;;;;;;;;;;;;;;;;;;;;;;;;;;;;;;;;;;;;;;;;;;;;;;;;;;;;;;;;;;;;;;;;;;;;;;;;;;;;

; Structure description for control latches.
(if (type (module $mod) clatch)
    (and (port data1 $mod) (input data1 $mod)
         (dsize (port data1 $mod) 256) (dsize (port data2 $mod) 16)
         (port data2 $mod) (output data2 $mod)
         (port clear $mod) (input clear $mod)
         (dsize (port clear $mod) 2) (dsize (port clock $mod) 2)
         (port clock $mod) (input clock $mod)
         (port irq $mod) (output irq $mod) (dsize (port irq $mod) 2)))

; Default preferences.
(if (type (module $mod) clatch) (bpref (module $mod) $x normal rule))

; Behavior for control latches.
(if (type (module $mod) clatch)
    (brules (module $mod) $x normal

; Define what it means for the clock to have a positive transition.
            (if (and (true (val (port clock $mod) (n 0 0)) $t)
                     (= $t (+ $t1 1))
                     (true (val (port clock $mod) (n 1 1)) $t1))
                (rising (val (port clock $mod) (n 1 1)) $t))

; If the clear signal is enabled, the output is all 0s.
           (if (and (true (val (port clear $mod) (n 1 1)) $t)
                    (= $t1 (+ $t 30)))
               (and (true (val (port data2 $mod)
                               (n (0 0 0 0) (0 0 0 0))) $t1)
                    (true (val (port irq   $mod) (n 0 0)) $t1)))

; Otherwise the output follows the input on rising clock transitions.
          (if (and (rising (val (port clock $mod) (n 1 1)) $t)
                   (true (val (port clear $mod) (n 0 0)) $t)
                   (true (val (port data1 $mod)
                              (n ($a $b $c $d $e $f $g $h)
                                 ($a $b $c $d $e $f $g $h)))
                         $t)
                   (= $t1 (+ $t 30)))
              (and (true (val (port irq $mod) (n $c $c)) $t1)
                   (true (val (port data2 $mod)
                              (n ($e $f $g $h)($e $f $g $h))) $t1)))

;define the "d" rules.
          (if (and (true (val (port data1 $mod)
                              (d ($a $b $c $d $e $f $g $h)
                                 ($i $j $k $l $m $n $o $p))) $t)
                   (bnot $c $k)
                   (rising (val (port clock $mod) (n 1 1)) $t)
                   (= $t1 (+ $t 30)))
              (true (val (port irq $mod) (d $c $k)) $t1))

          (if (and (true (val (port data1 $mod)
                              (d ($a $b $c $d $e $f $g $h)
                                 ($i $j $k $l $m $n $o $p))) $t)
                   (nmatch ($e $f $g $h) ($m $n $o $p));nibbles diff.
                   (rising (val (port clock $mod) (n 1 1)) $t)
                   (= $t1 (+ $t 30)))
              (true (val (port data2 $mod)
                         (d ($e $f $g $h) ($m $n $o $p))) $t1))
```

```
        (if (and (true (val (port clear $mod) (d $a $b)) $t)
                 (true (val (port irq $mod) (n 1 1)) $t)
                 (= $t1 (+ $t 30)))
            (true (val (port irq $mod) (d $b $a)) $t1))

        (if (and (true (val (port clear $mod) (d $a $b)) $t)
                 (true (val (port data2 $mod)
                            (n (1 1 1 1) (1 1 1 1))) $t)
                 (= $t1 (+ $t 30)))
            (true (val (port data2 $mod)
                       (d ($b $b $b $b) ($a $a $a $a))) $t1))

        (if (and (true (val (port clock $mod) (d $a $b)) $t)
                 (= $t (+ $t1 40))
                 (true (val (port clock $mod)
                            (n 0 0)) $t1) ; clock was 0.
                 (true (val (port clear $mod)
                            (n 1 1)) $t1); reset outputs.
                 (true (val (port data1 $mod)
                            (n (1 1 1 1 1 1 1 1)
                               (1 1 1 1 1 1 1 1))) $t1)
                 (= $t2 (+ $t 30)))
            (and (true (val (port irq $mod) (d $a $b)) $t2)
                 (true (val (port data2 $mod)
                            (d ($a $a $a $a)
                               ($b $b $b $b))) $t2)))
))
```

```
;;;;;;;;;;;;;;;;;;;;;;;;;;;;;;;;;;;;;;;;;;;;;;;;;;;;;;;;;;;;;;;;;;;;;;;;
;;;              Define how to test the control latch                 ;;;
;;;;;;;;;;;;;;;;;;;;;;;;;;;;;;;;;;;;;;;;;;;;;;;;;;;;;;;;;;;;;;;;;;;;;;;;

(if (type (module $mod) clatch)
    (trules (module $mod) behavior ; if testing via top-level.

; test the clear signal.
           (if (true (val (port clear $mod) (n 1 1)) 0)
               (true (val (port data2 $mod) (d (0 0 0 0)
                                               (1 1 1 1))) 30))

           (if (true (val (port clear $mod) (n 1 1)) 0)
               (true (val (port irq   $mod) (d 0 1)) 30))

; test the output for rising clock transitions
          (if (and (true (val (port clock $mod) (n 0 0)) 0)
                   (true (val (port clock $mod) (n 1 1)) 1)
                   (true (val (port clear $mod) (n 0 0)) 0)
                   (true (val (port data1 $mod)
                              (n (1 0 1 0 1 0 1 0)
                                 (1 0 1 0 1 0 1 0))) 0))
              (true (val (port data2 $mod) (d (1 0 1 0)
                                              (1 1 1 1))) 30))

          (if (and (true (val (port clock $mod) (n 0 0)) 0)
                   (true (val (port clock $mod) (n 1 1)) 1)
                   (true (val (port clear $mod) (n 0 0)) 0)
                   (true (val (port data1 $mod)
                              (n (0 1 0 1 0 1 0 1)
                                 (0 1 0 1 0 1 0 1))) 0))
              (true (val (port data2 $mod) (d (1 0 1 0)
                                              (1 1 1 1))) 30))

          (if (and (true (val (port clock $mod) (n 0 0)) 0)
                   (true (val (port clock $mod) (n 1 1)) 1)
                   (true (val (port clear $mod) (n 0 0)) 0)
                   (true (val (port data1 $mod)
                              (n (1 0 1 0 1 0 1 0)
                                 (1 0 1 0 1 0 1 0))) 0))
              (true (val (port irq $mod) (d 1 0)) 30))

          (if (and (true (val (port clock $mod) (n 0 0)) 0)
                   (true (val (port clock $mod) (n 1 1)) 1)
                   (true (val (port clear $mod) (n 0 0)) 0)
                   (true (val (port data1 $mod)
                              (n (0 1 0 1 0 1 0 1)
                                 (0 1 0 1 0 1 0 1))) 0))
              (true (val (port irq $mod) (d 0 1)) 30)) ))
```

```
;;;;;;;;;;;;;;;;;;;;;;;;;;;;;;;;;;;;;;;;;;;;;;;;;;;;;;;;;;;;;;;;;;;;;;;;;;;
;;;          Corona definitions for the data latch prototype          ;;;
;;;;;;;;;;;;;;;;;;;;;;;;;;;;;;;;;;;;;;;;;;;;;;;;;;;;;;;;;;;;;;;;;;;;;;;;;;;

; Structure description for data latches.
(if (type (module $mod) dlatch)
    (and (port data1 $mod) (input data1 $mod)
         (dsize (port data1 $mod) 256) (dsize (port data2 $mod) 256)
         (port data2 $mod) (output data2 $mod)
         (port clock $mod) (input clock $mod)
         (dsize (port clock $mod) 2)))

; Default preferences.
(if (type (module $mod) dlatch) (bpref (module $mod) $x normal rule))

; Behavior for data latches.
(if (type (module $mod) dlatch)
    (brules (module $mod) $x normal

; Define what it means for the clock to have a positive transition.
            (if (and (true (val (port clock $mod) (n 0 0)) $t)
                     (= $t (+ $t1 1))
                     (true (val (port clock $mod) (n 1 1)) $t1))
                (rising (val (port clock $mod) (n 1 1)) $t))

; If the clock is rising, copy the input to the output.
            (if (and (rising (val (port clock $mod) (n 1 1)) $t)
                     (true (val (port data1 $mod) $v) $t)
                     (= $t1 (+ $t 30)))
                (true (val (port data2 $mod) $v) $t1))

; specify "d" rule for the clock input.
            (if (and (true (val (port clock $mod) (d $a $b)) $t)
                     (= $t (+ $t1 1))
                     (true (val (port data1 $mod)
                                (n (0 0 0 0 0 0 0 0)
                                   (0 0 0 0 0 0 0 0))) $t1)
                     (true (val (port clock $mod) (n 0 0)) $t1)
                     (= $t2 (+ $t 30)))
                (true (val (port data2 $mod)
                           (d ($b $b $b $b $b $b $b $b)
                              ($a $a $a $a $a $a $a $a))) $t2))

))
```

```
;;;;;;;;;;;;;;;;;;;;;;;;;;;;;;;;;;;;;;;;;;;;;;;;;;;;;;;;;;;;;;;;;;;;;;;;;;;;;;
;;;              Define how to test the data latch                        ;;;
;;;;;;;;;;;;;;;;;;;;;;;;;;;;;;;;;;;;;;;;;;;;;;;;;;;;;;;;;;;;;;;;;;;;;;;;;;;;;;

(if (type (module $mod) dlatch)
    (trules (module $mod) behavior ; if testing via top-level.

; check the output for a rising clock input.
            (if (and (true (val (port clock $mod) (n 0 0)) 0)
                     (true (val (port clock $mod) (n 1 1)) 1)
                     (true (val (port data1 $mod)
                                 (n (1 0 1 0 1 0 1 0)
                                    (1 0 1 0 1 0 1 0))) 0))
                (true (val (port data2 $mod)
                           (d (1 0 1 0 1 0 1 0)
                              (1 1 1 1 1 1 1 1))) 30))

            (if (and (true (val (port clock $mod) (n 0 0)) 0)
                     (true (val (port clock $mod) (n 1 1)) 1)
                     (true (val (port data1 $mod)
                                 (n (0 1 0 1 0 1 0 1)
                                    (0 1 0 1 0 1 0 1))) 0))
                (true (val (port data2 $mod)
                           (d (0 1 0 1 0 1 0 1)
                              (1 1 1 1 1 1 1 1))) 30))

))
```

```
;;;;;;;;;;;;;;;;;;;;;;;;;;;;;;;;;;;;;;;;;;;;;;;;;;;;;;;;;;;;;;;;;;;;;;;;;;
;;;          Corona definitions for the driver prototype             ;;;
;;;;;;;;;;;;;;;;;;;;;;;;;;;;;;;;;;;;;;;;;;;;;;;;;;;;;;;;;;;;;;;;;;;;;;;;;;

; Structure description for control drivers.
(if (type (module $mod) driver)
    (and (port data $mod) (input data $mod) (dsize (port data $mod) 16)
         (port control $mod) (output control $mod)
         (dsize (port control $mod) 16)))

; Default preferences.
(if (type (module $mod) driver) (bpref (module $mod) $x normal rule))

; Behavior for the control drivers.
(if (type (module $mod) driver)
    (brules (module $mod) $x normal

;  Output follows input with almost complete inversion.
            (if (and (true (val (port data $mod)
                                (n ($a $b $c $d)
                                   ($a $b $c $d))) $t)
                     (bnot $d $di)
                     (bnot $c $ci)
                     (bnot $a $ai)
                     (= $t1 (+ $t 30)))
                (true (val (port control $mod)
                           (n ($ai $b $ci $di)
                              ($ai $b $ci $di))) $t1))

; define the "d" rules.
            (if (and (true (val (port data $mod)
                                (d ($a $b $c $d)
                                   ($e $f $g $h))) $t)
                     (bnot $d $di)
                     (bnot $h $hi)
                     (bnot $c $ci)
                     (bnot $g $gi)
                     (bnot $a $ai)
                     (bnot $e $ei)
                     (= $t1 (+ $t 30)))
                (true (val (port control $mod)
                           (d ($ai $b $ci $di)
                              ($ei $f $gi $hi))) $t1))

))
```

```
;;;;;;;;;;;;;;;;;;;;;;;;;;;;;;;;;;;;;;;;;;;;;;;;;;;;;;;;;;;;;;;;;;;;;;;;;;;;;;
;;;                    Define how to test the driver                       ;;;
;;;;;;;;;;;;;;;;;;;;;;;;;;;;;;;;;;;;;;;;;;;;;;;;;;;;;;;;;;;;;;;;;;;;;;;;;;;;;;

(if (type (module $mod) driver)
    (trules (module $mod) behavior ; if testing via top-level.

;  test the output for a couple of inputs.
            (if (true (val (port data $mod) (n (1 0 1 0) (1 0 1 0))) 0)
                (true (val (port control $mod) (d (0 0 0 1)
                                                   (1 1 1 1))) 30))

            (if (true (val (port data $mod) (n (0 1 0 1) (0 1 0 1))) 0)
                (true (val (port control $mod) (d (1 1 1 0)
                                                   (1 1 1 1))) 30))
))
```

```
;;;;;;;;;;;;;;;;;;;;;;;;;;;;;;;;;;;;;;;;;;;;;;;;;;;;;;;;;;;;;;;;;;;;;;;;
;;;            Corona definitions for the local-bus prototype        ;;;
;;;;;;;;;;;;;;;;;;;;;;;;;;;;;;;;;;;;;;;;;;;;;;;;;;;;;;;;;;;;;;;;;;;;;;;;

; Define structure for local buses.
(if (and (type (module $mod) (lbus $isize))
         (range $h 1 $isize))
    (and (port $h $mod) (input $h $mod) (output $h $mod)
         (dsize (port $h $mod) 255)))

; Default preferences.
(if (type (module $mod) (lbus $isize))
    (bpref (module $mod) $x normal rule))

; Define behavior for local buses- assumes at most one driver.
(if (type (module $mod) (lbus $isize))
    (brules (module $mod) $x normal

; if a port is set to a value, set all ports to that value, assume that
; this is the only driver. This rule covers both "d" and "n" values.
            (if (and (true (val (port $x $mod) $v) $t);any port is set.
                     (port $f $mod) ; find all ports.
                     (= $t1 (+ $t 10))) ;10ns delay for bus settling.
                (true (val (port $f $mod) $v) $t1))

; this rule handles backward propagation.
            (if (and (true (val (port $x $mod) $v) $t);any port is set.
                     (port $x $mod) ; find all ports.
                     (= $t1 (+ $t 10))) ;10ns delay for bus settling.
                (true (val (port $f $mod) $v) $t1))

; need to prevent infinite loop via control mechanism.

))
```

Appendix B

Tests for the Printer Adapter Card

This appendix defines the tests for the IBM PC Printer Adapter card. These tests were generated by the Saturn test generation system using the design descriptions presented in the previous appendix. The tests took approximately 30 minutes to generate on a PDP-20 running MRS version 7.1 [24] under Maclisp [46]. This time includes the time required to pretty print the trace of the test generation process.

To minimize the number of tests the card is tested by testing each component (cdecoder, transceiver, etc.) individually. Part of the design specification includes a description of how to test each component. Each test for a component specifies a sequence of values for the input ports and the expected value of an output port for these inputs. In this appendix each test is prefixed by the string `Trying to achieve test:`, and followed by a solution to this test. The solution of the test specifies the values for the inputs of the Printer Adapter card and the expected value for an output port of this card for these inputs. These results were obtained from a *photo* recording session of the system.

```
Generating tests for the command decoder:

Trying to achieve test:

(IF (AND (TRUE (VAL (PORT ADDRESS PA-DECODER)
                    (N (1 1 0 1 1 1 1 $222 0 0)
                       (1 1 0 1 1 1 1 $222 0 0)))
               0)
         (TRUE (VAL (PORT CONTROL PA-DECODER)
                    (N ($223 $224 0 $225 $226 $227)
                       ($223 $224 0 $225 $226 $227)))
               0))
    (TRUE (VAL (PORT RDATA PA-DECODER) (D 0 1)) 100))

Solution:

(IF (AND (TRUE (VAL (PORT BUS-CONTROL PA)
                    (N ($15 $16 0 $17 $18 $19)
                       ($15 $16 0 $17 $18 $19)))
               10)
         (TRUE (VAL (PORT BUS-ADDRESS PA)
                    (N (1 1 0 1 1 1 1 $8 $9 $10)
                       (1 1 0 1 1 1 1 $8 $9 $10)))
               10)
         (TRUE (VAL (PORT BUS-ADDRESS PA)
                    (N ($12 $2 $3 $4 $5 $6 0 $8 $9 $10)
                       ($12 $2 $3 $4 $5 $6 0 $8 $9 $10)))
               -30)
         (TRUE (VAL (PORT BUS-ADDRESS PA)
                    (N (1 1 0 1 1 1 1 $8 0 0)
                       (1 1 0 1 1 1 1 $8 0 0)))
               -31)
         (TRUE (VAL (PORT BUS-CONTROL PA)
                    (N ($15 $16 $20 0 $18 $19)
                       ($15 $16 $20 0 $18 $19)))
               -31)
         (TRUE (VAL (PORT BUS-ADDRESS PA)
                    (N ($12 $2 $3 $4 $5 $6 0 $8 $9 $10)
                       ($12 $2 $3 $4 $5 $6 0 $8 $9 $10)))
               -80)
         (TRUE (VAL (PORT BUS-DATA PA)
                    (N (1 0 1 0 1 0 1 0) (1 0 1 0 1 0 1 0)))
               20)
         (TRUE (VAL (PORT BUS-ADDRESS PA)
                    (N (1 1 0 1 1 1 1 $222 0 0)
                       (1 1 0 1 1 1 1 $222 0 0)))
               0)
         (TRUE (VAL (PORT BUS-CONTROL PA)
                    (N ($223 $224 0 $225 $226 $227)
                       ($223 $224 0 $225 $226 $227)))
               0))
    (TRUE (VAL (PORT BUS-DATA PA)
               (D (1 0 1 0 1 0 1 0) (1 1 1 1 1 1 1 1)))
          140))

Trying to achieve test:

(IF (TRUE (VAL (PORT CONTROL PA-DECODER)
               (N ($223 $224 1 $225 $226 $227)
                  ($223 $224 1 $225 $226 $227)))
          0)
    (TRUE (VAL (PORT RDATA PA-DECODER) (D 1 0)) 100))
```

```
Solution:

(IF (AND (TRUE (VAL (PORT BUS-ADDRESS PA)
                    (N ($12 $2 $3 $4 $5 $6 0 $8 $9 $10)
                       ($12 $2 $3 $4 $5 $6 0 $8 $9 $10)))
               -30)
         (TRUE (VAL (PORT BUS-ADDRESS PA)
                    (N (1 1 0 1 1 1 1 $8 0 0)
                       (1 1 0 1 1 1 1 $8 0 0)))
               -31)
         (TRUE (VAL (PORT BUS-CONTROL PA)
                    (N ($15 $16 $20 0 $18 $19)
                       ($15 $16 $20 0 $18 $19)))
               -31)
         (TRUE (VAL (PORT BUS-ADDRESS PA)
                    (N ($12 $2 $3 $4 $5 $6 0 $8 $9 $10)
                       ($12 $2 $3 $4 $5 $6 0 $8 $9 $10)))
               -80)
         (TRUE (VAL (PORT BUS-DATA PA)
                    (N (1 0 1 0 1 0 1 0) (1 0 1 0 1 0 1 0)))
               20)
         (TRUE (VAL (PORT BUS-CONTROL PA)
                    (N ($15 $16 0 $17 $18 $19)
                       ($15 $16 0 $17 $18 $19)))
               30)
         (TRUE (VAL (PORT BUS-ADDRESS PA)
                    (N (1 1 0 1 1 1 1 $8 $9 $10)
                       (1 1 0 1 1 1 1 $8 $9 $10)))
               30)
         (TRUE (VAL (PORT BUS-CONTROL PA)
                    (N ($223 $224 1 $225 $226 $227)
                       ($223 $224 1 $225 $226 $227)))
               0))
    (TRUE (VAL (PORT BUS-DATA PA)
               (D (1 1 1 1 1 1 1 1) (1 0 1 0 1 0 1 0)))
          160))

Trying to achieve test:

(IF (TRUE (VAL (PORT ADDRESS PA-DECODER)
               (N ($228 $229 $230 $231 $232 $233 $234 $222 $235 1)
                  ($228 $229 $230 $231 $232 $233 $234 $222 $235 1)))
          0)
    (TRUE (VAL (PORT RDATA PA-DECODER) (D 1 0)) 100))

Solution:

(IF (AND (TRUE (VAL (PORT BUS-ADDRESS PA)
                    (N ($12 $2 $3 $4 $5 $6 0 $8 $9 $10)
                       ($12 $2 $3 $4 $5 $6 0 $8 $9 $10)))
               -30)
         (TRUE (VAL (PORT BUS-ADDRESS PA)
                    (N (1 1 0 1 1 1 1 $8 0 0)
                       (1 1 0 1 1 1 1 $8 0 0)))
               -31)
         (TRUE (VAL (PORT BUS-CONTROL PA)
                    (N ($15 $16 $20 0 $18 $19)
                       ($15 $16 $20 0 $18 $19)))
               -31)
         (TRUE (VAL (PORT BUS-ADDRESS PA)
                    (N ($12 $2 $3 $4 $5 $6 0 $8 $9 $10)
                       ($12 $2 $3 $4 $5 $6 0 $8 $9 $10)))
               -80)
```

```
        (TRUE (VAL (PORT BUS-DATA PA)
                   (N (1 0 1 0 1 0 1 0) (1 0 1 0 1 0 1 0)))
              20)
        (TRUE (VAL (PORT BUS-CONTROL PA)
                   (N ($15 $16 0 $17 $18 $19)
                      ($15 $16 0 $17 $18 $19)))
              30)
        (TRUE (VAL (PORT BUS-ADDRESS PA)
                   (N (1 1 0 1 1 1 1 $8 $9 $10)
                      (1 1 0 1 1 1 1 $8 $9 $10)))
              30)
        (TRUE (VAL (PORT BUS-ADDRESS PA)
                   (N ($228 $229 $230 $231 $232 $233 $234 $222 $235 1)
                      ($228 $229 $230 $231 $232 $233 $234 $222 $235 1)))
              0))
    (TRUE (VAL (PORT BUS-DATA PA)
               (D (1 1 1 1 1 1 1 1) (1 0 1 0 1 0 1 0)))
          160))
```

```
Trying to achieve test:

(IF (AND (TRUE (VAL (PORT ADDRESS PA-DECODER)
                    (N (1 1 0 1 1 1 1 $222 0 1)
                       (1 1 0 1 1 1 1 $222 0 1)))
               0)
         (TRUE (VAL (PORT CONTROL PA-DECODER)
                    (N ($223 $224 0 $225 $226 $227)
                       ($223 $224 0 $225 $226 $227)))
               0))
    (TRUE (VAL (PORT RSTATUS PA-DECODER) (D 0 1)) 100))

Solution:

(IF (AND (TRUE (VAL (PORT PBUS-STATUS PA) (N (1 0 0 0 0) (1 0 0 0 0)))
               100)
         (TRUE (VAL (PORT BUS-ADDRESS PA)
                    (N (1 1 0 1 1 1 1 $8 $9 $10)
                       (1 1 0 1 1 1 1 $8 $9 $10)))
               80)
         (TRUE (VAL (PORT BUS-CONTROL PA)
                    (N ($15 $16 0 $17 $18 $19)
                       ($15 $16 0 $17 $18 $19)))
               80)
         (TRUE (VAL (PORT BUS-ADDRESS PA)
                    (N ($228 $229 $230 $231 $232 $233 $234 $222 $235 1)
                       ($228 $229 $230 $231 $232 $233 $234 $222 $235 1)))
               0)
         (TRUE (VAL (PORT BUS-CONTROL PA)
                    (N ($223 $224 0 $225 $226 $227)
                       ($223 $224 0 $225 $226 $227)))
               0))
    (TRUE (VAL (PORT BUS-DATA PA)
               (D (0 0 0 0 0 1 1 1) (1 1 1 1 1 1 1 1)))
          210))

Trying to achieve test:

(IF (TRUE (VAL (PORT CONTROL PA-DECODER)
               (N ($223 $224 1 $225 $226 $227)
                  ($223 $224 1 $225 $226 $227)))
          0)
    (TRUE (VAL (PORT RSTATUS PA-DECODER) (D 1 0)) 100))
```

```
Solution:

(IF (AND (TRUE (VAL (PORT PBUS-STATUS PA) (N (1 0 0 0 0) (1 0 0 0 0)))
              100)
         (TRUE (VAL (PORT BUS-ADDRESS PA)
                    (N (1 1 0 1 1 1 1 $8 $9 $10)
                       (1 1 0 1 1 1 1 $8 $9 $10)))
              80)
         (TRUE (VAL (PORT BUS-CONTROL PA)
                    (N ($15 $16 0 $17 $18 $19)
                       ($15 $16 0 $17 $18 $19)))
              80)
         (TRUE (VAL (PORT BUS-CONTROL PA)
                    (N ($223 $224 1 $225 $226 $227)
                       ($223 $224 1 $225 $226 $227)))
              0))
    (TRUE (VAL (PORT BUS-DATA PA)
               (D (1 1 1 1 1 1 1 1) (0 0 0 0 0 1 1 1)))
          210))

Trying to achieve test:

(IF (TRUE (VAL (PORT ADDRESS PA-DECODER)
               (N ($228 $229 $230 $231 $232 $233 $234 $222 1 $236)
                  ($228 $229 $230 $231 $232 $233 $234 $222 1 $236)))
          0)
    (TRUE (VAL (PORT RSTATUS PA-DECODER) (D 1 0)) 100))

Solution:

(IF (AND (TRUE (VAL (PORT PBUS-STATUS PA) (N (1 0 0 0 0) (1 0 0 0 0)))
              100)
         (TRUE (VAL (PORT BUS-ADDRESS PA)
                    (N (1 1 0 1 1 1 1 $8 $9 $10)
                       (1 1 0 1 1 1 1 $8 $9 $10)))
              80)
         (TRUE (VAL (PORT BUS-CONTROL PA)
                    (N ($15 $16 0 $17 $18 $19)
                       ($15 $16 0 $17 $18 $19)))
              80)
         (TRUE (VAL (PORT BUS-ADDRESS PA)
                    (N (1 1 0 1 1 1 1 $222 1 0)
                       (1 1 0 1 1 1 1 $222 1 0)))
              0))
    (TRUE (VAL (PORT BUS-DATA PA)
               (D (1 1 1 1 1 1 1 1) (0 0 0 0 0 1 1 1)))
          210))

Trying to achieve test:

(IF (AND (TRUE (VAL (PORT ADDRESS PA-DECODER)
                    (N (1 1 0 1 1 1 1 $222 1 0)
                       (1 1 0 1 1 1 1 $222 1 0)))
              0)
         (TRUE (VAL (PORT CONTROL PA-DECODER)
                    (N ($223 $224 0 $225 $226 $227)
                       ($223 $224 0 $225 $226 $227)))
              0))
    (TRUE (VAL (PORT RCONTROL PA-DECODER) (D 0 1)) 100))

Solution:

(IF (AND (TRUE (VAL (PORT BUS-CONTROL PA)
```

```
                 (N ($15 $16 0 $17 $18 $19)
                    ($15 $16 0 $17 $18 $19)))
             70)
        (TRUE (VAL (PORT BUS-ADDRESS PA)
                   (N (1 1 0 1 1 1 1 $8 $9 $10)
                      (1 1 0 1 1 1 1 $8 $9 $10)))
              70)
        (TRUE (VAL (PORT BUS-RESET PA) (N 1 1)) 70)
        (TRUE (VAL (PORT BUS-RESET PA) (N 1 1)) 40)
        (TRUE (VAL (PORT BUS-ADDRESS PA)
                   (N (1 1 0 1 1 1 1 $222 1 0)
                      (1 1 0 1 1 1 1 $222 1 0)))
              0)
        (TRUE (VAL (PORT BUS-CONTROL PA)
                   (N ($223 $224 0 $225 $226 $227)
                      ($223 $224 0 $225 $226 $227)))
              0))
    (TRUE (VAL (PORT BUS-DATA PA)
               (D (1 1 1 0 0 0 0 0) (1 1 1 1 1 1 1 1)))
          200))
```

Trying to achieve test:

```
(IF (TRUE (VAL (PORT CONTROL PA-DECODER)
               (N ($223 $224 1 $225 $226 $227)
                  ($223 $224 1 $225 $226 $227)))
          0)
    (TRUE (VAL (PORT RCONTROL PA-DECODER) (D 1 0)) 100))
```

Solution:

```
(IF (AND (TRUE (VAL (PORT BUS-RESET PA) (N 1 1)) 70)
         (TRUE (VAL (PORT BUS-RESET PA) (N 1 1)) 40)
         (TRUE (VAL (PORT BUS-ADDRESS PA)
                    (N (1 1 0 1 1 1 1 $8 $9 $10)
                       (1 1 0 1 1 1 1 $8 $9 $10)))
               80)
         (TRUE (VAL (PORT BUS-CONTROL PA)
                    (N ($15 $16 0 $17 $18 $19)
                       ($15 $16 0 $17 $18 $19)))
               80)
         (TRUE (VAL (PORT BUS-CONTROL PA)
                    (N ($223 $224 1 $225 $226 $227)
                       ($223 $224 1 $225 $226 $227)))
               0))
    (TRUE (VAL (PORT BUS-DATA PA)
               (D (1 1 1 1 1 1 1 1) (1 1 1 0 0 0 0 0)))
          210))
```

Trying to achieve test:

```
(IF (TRUE (VAL (PORT ADDRESS PA-DECODER)
               (N ($228 $229 $230 $231 $232 $233 $234 $222 $235 1)
                  ($228 $229 $230 $231 $232 $233 $234 $222 $235 1)))
          0)
    (TRUE (VAL (PORT RCONTROL PA-DECODER) (D 1 0)) 100))
```

Solution:

```
(IF (AND (TRUE (VAL (PORT BUS-RESET PA) (N 1 1)) 70)
         (TRUE (VAL (PORT BUS-RESET PA) (N 1 1)) 40)
         (TRUE (VAL (PORT BUS-ADDRESS PA)
                    (N (1 1 0 1 1 1 1 $8 $9 $10)
```

```
                    (1 1 0 1 1 1 1 $8 $9 $10)))
                80)
           (TRUE (VAL (PORT BUS-CONTROL PA)
                    (N ($15 $16 0 $17 $18 $19)
                       ($15 $16 0 $17 $18 $19)))
                80)
           (TRUE (VAL (PORT BUS-ADDRESS PA)
                    (N ($228 $229 $230 $231 $232 $233 $234 $222 $235 1)
                       ($228 $229 $230 $231 $232 $233 $234 $222 $235 1)))
                0))
       (TRUE (VAL (PORT BUS-DATA PA)
                (D (1 1 1 1 1 1 1 1) (1 1 1 0 0 0 0 0)))
             210))
```

Trying to achieve test:

```
(IF (AND (TRUE (VAL (PORT ADDRESS PA-DECODER)
                    (N (1 1 0 1 1 1 1 $222 $235 $236)
                       (1 1 0 1 1 1 1 $222 $235 $236)))
                0)
           (TRUE (VAL (PORT CONTROL PA-DECODER)
                    (N ($223 $224 0 $225 $226 $227)
                       ($223 $224 0 $225 $226 $227)))
                0))
       (TRUE (VAL (PORT DIR PA-DECODER) (D 0 1)) 100))
```

Solution:

```
(IF (AND (TRUE (VAL (PORT BUS-ADDRESS PA)
                    (N ($12 $2 $3 $4 $5 $6 0 $8 $9 $10)
                       ($12 $2 $3 $4 $5 $6 0 $8 $9 $10)))
                -30)
           (TRUE (VAL (PORT BUS-DATA PA)
                    (N (1 0 1 0 1 0 1 0) (1 0 1 0 1 0 1 0)))
                70)
           (TRUE (VAL (PORT BUS-ADDRESS PA)
                    (N (1 1 0 1 1 1 1 $222 1 0)
                       (1 1 0 1 1 1 1 $222 1 0)))
                0)
           (TRUE (VAL (PORT BUS-CONTROL PA)
                    (N ($223 $224 0 $225 $226 $227)
                       ($223 $224 0 $225 $226 $227)))
                0))
       (TRUE (VAL (PORT BUS-DATA PA)
                (D (1 0 1 0 1 0 1 0) (1 1 1 1 1 1 1 1)))
             130))
```

Trying to achieve test:

```
(IF (TRUE (VAL (PORT CONTROL PA-DECODER)
               (N ($223 $224 1 $225 $226 $227)
                  ($223 $224 1 $225 $226 $227)))
           0)
       (TRUE (VAL (PORT DIR PA-DECODER) (D 1 0)) 100))
```

Solution:

```
(IF (AND (TRUE (VAL (PORT BUS-ADDRESS PA)
                    (N (1 1 0 1 1 1 1 $8 $9 $10)
                       (1 1 0 1 1 1 1 $8 $9 $10)))
                30)
           (TRUE (VAL (PORT BUS-CONTROL PA)
                    (N ($15 $16 0 $17 $18 $19)
```

```
                    ($15 $16 0 $17 $18 $19)))
               30)
         (TRUE (VAL (PORT BUS-DATA PA)
                    (N (1 0 1 0 1 0 1 0) (1 0 1 0 1 0 1 0)))
               100)
         (TRUE (VAL (PORT BUS-CONTROL PA)
                    (N ($223 $224 1 $225 $226 $227)
                       ($223 $224 1 $225 $226 $227)))
               0))
    (TRUE (VAL (PORT BUS-DATA PA)
               (D (1 0 1 0 1 0 1 0) (1 1 1 1 1 1 1 1)))
          160))
```

```
Trying to achieve test:
```

```
(IF (TRUE (VAL (PORT ADDRESS PA-DECODER)
               (N (0 1 0 1 1 1 1 $22 $35 $36)
                  (0 1 0 1 1 1 1 $22 $35 $36)))
          0)
    (TRUE (VAL (PORT DIR PA-DECODER) (D 1 0)) 100))
```

```
Solution:
```

```
(IF (AND (TRUE (VAL (PORT BUS-ADDRESS PA)
                    (N (1 1 0 1 1 1 1 $8 $9 $10)
                       (1 1 0 1 1 1 1 $8 $9 $10)))
               30)
         (TRUE (VAL (PORT BUS-CONTROL PA)
                    (N ($15 $16 0 $17 $18 $19)
                       ($15 $16 0 $17 $18 $19)))
               30)
         (TRUE (VAL (PORT BUS-DATA PA)
                    (N (1 0 1 0 1 0 1 0) (1 0 1 0 1 0 1 0)))
               100)
         (TRUE (VAL (PORT BUS-ADDRESS PA)
                    (N ($28 $29 $30 $31 $32 $33 $34 $22 $35 1)
                       ($28 $29 $30 $31 $32 $33 $34 $22 $35 1)))
               0))
    (TRUE (VAL (PORT BUS-DATA PA)
               (D (1 0 1 0 1 0 1 0) (1 1 1 1 1 1 1 1)))
          160))
```

```
Trying to achieve test:
```

```
(IF (AND (TRUE (VAL (PORT ADDRESS PA-DECODER)
                    (N (1 1 0 1 1 1 1 $222 0 0)
                       (1 1 0 1 1 1 1 $222 0 0)))
               0)
         (TRUE (VAL (PORT CONTROL PA-DECODER)
                    (N ($223 $224 $237 0 $226 $227)
                       ($223 $224 $237 0 $226 $227)))
               0))
    (TRUE (VAL (PORT WDATA PA-DECODER) (D 0 1)) 100))
```

```
Solution:
```

```
(IF (AND (TRUE (VAL (PORT BUS-CONTROL PA)
                    (N ($15 $16 $20 0 $18 $19)
                       ($15 $16 $20 0 $18 $19)))
               -1)
         (TRUE (VAL (PORT BUS-ADDRESS PA)
                    (N (1 1 0 1 1 1 1 $8 0 0)
                       (1 1 0 1 1 1 1 $8 0 0)))
```

```
              -1)
          (TRUE (VAL (PORT BUS-DATA PA)
                     (N (0 0 0 0 0 0 0 0) (0 0 0 0 0 0 0 0)))
              49)
          (TRUE (VAL (PORT BUS-ADDRESS PA)
                     (N ($12 $2 $3 $4 $5 0 $7 $8 $9 $10)
                        ($12 $2 $3 $4 $5 0 $7 $8 $9 $10)))
              -51)
          (TRUE (VAL (PORT BUS-ADDRESS PA)
                     (N (1 1 0 1 1 1 1 $222 0 0)
                        (1 1 0 1 1 1 1 $222 0 0)))
              0)
          (TRUE (VAL (PORT BUS-CONTROL PA)
                     (N ($223 $224 1 $225 $226 $227)
                        ($223 $224 1 $225 $226 $227)))
              0))
    (TRUE (VAL (PORT PBUS-DATA PA)
               (D (1 1 1 1 1 1 1 1) (0 0 0 0 0 0 0 0)))
          130))
```

Trying to achieve test:

```
(IF (TRUE (VAL (PORT CONTROL PA-DECODER)
               (N ($223 $224 $237 1 $226 $227)
                  ($223 $224 $237 1 $226 $227)))
          0)
    (TRUE (VAL (PORT WDATA PA-DECODER) (D 1 0)) 100))
```

Solution:

```
(IF (AND (TRUE (VAL (PORT BUS-CONTROL PA)
                    (N ($15 $16 $20 0 $18 $19)
                       ($15 $16 $20 0 $18 $19)))
              -1)
         (TRUE (VAL (PORT BUS-ADDRESS PA)
                    (N (1 1 0 1 1 1 1 $8 0 0)
                       (1 1 0 1 1 1 1 $8 0 0)))
              -1)
         (TRUE (VAL (PORT BUS-DATA PA)
                    (N (0 0 0 0 0 0 0 0) (0 0 0 0 0 0 0 0)))
              49)
         (TRUE (VAL (PORT BUS-ADDRESS PA)
                    (N ($12 $2 $3 $4 $5 0 $7 $8 $9 $10)
                       ($12 $2 $3 $4 $5 0 $7 $8 $9 $10)))
              -51)
         (TRUE (VAL (PORT BUS-CONTROL PA)
                    (N ($223 $224 1 $225 $226 $227)
                       ($223 $224 1 $225 $226 $227)))
              0))
    (TRUE (VAL (PORT PBUS-DATA PA)
               (D (0 0 0 0 0 0 0 0) (1 1 1 1 1 1 1 1)))
          130))
```

Trying to achieve test:

```
(IF (TRUE (VAL (PORT ADDRESS PA-DECODER)
               (N ($228 $229 $230 $231 $232 $233 $234 $222 1 $236)
                  ($228 $229 $230 $231 $232 $233 $234 $222 1 $236)))
          0)
    (TRUE (VAL (PORT WDATA PA-DECODER) (D 1 0)) 100))
```

Solution:

```
(IF (AND (TRUE (VAL (PORT BUS-CONTROL PA)
                    (N ($15 $16 $20 0 $18 $19)
                       ($15 $16 $20 0 $18 $19)))
                -1)
         (TRUE (VAL (PORT BUS-ADDRESS PA)
                    (N (1 1 0 1 1 1 1 $8 0 0)
                       (1 1 0 1 1 1 1 $8 0 0)))
                -1)
         (TRUE (VAL (PORT BUS-DATA PA)
                    (N (0 0 0 0 0 0 0 0) (0 0 0 0 0 0 0 0)))
                49)
         (TRUE (VAL (PORT BUS-ADDRESS PA)
                    (N ($12 $2 $3 $4 $5 0 $7 $8 $9 $10)
                       ($12 $2 $3 $4 $5 0 $7 $8 $9 $10)))
                -51)
         (TRUE (VAL (PORT BUS-ADDRESS PA)
                    (N (1 1 0 1 1 1 1 $222 1 0)
                       (1 1 0 1 1 1 1 $222 1 0)))
                0))
    (TRUE (VAL (PORT PBUS-DATA PA)
                (D (0 0 0 0 0 0 0 0) (1 1 1 1 1 1 1 1)))
           130))
```

Trying to achieve test:

```
(IF (AND (TRUE (VAL (PORT ADDRESS PA-DECODER)
                    (N (1 1 0 1 1 1 1 $222 1 0)
                       (1 1 0 1 1 1 1 $222 1 0)))
                0)
         (TRUE (VAL (PORT CONTROL PA-DECODER)
                    (N ($223 $224 $237 0 $226 $227)
                       ($223 $224 $237 0 $226 $227)))
                0))
    (TRUE (VAL (PORT WCONTROL PA-DECODER) (D 0 1)) 100))
```

Solution:

```
(IF (AND (TRUE (VAL (PORT BUS-RESET PA) (N 1 1)) 60)
         (TRUE (VAL (PORT BUS-CONTROL PA)
                    (N ($15 $16 $20 0 $18 $19)
                       ($15 $16 $20 0 $18 $19)))
                -40)
         (TRUE (VAL (PORT BUS-ADDRESS PA)
                    (N (1 1 0 1 1 1 1 $8 1 0)
                       (1 1 0 1 1 1 1 $8 1 0)))
                -40)
         (TRUE (VAL (PORT BUS-ADDRESS PA)
                    (N ($12 $2 $3 $4 $5 $6 0 $8 $9 $10)
                       ($12 $2 $3 $4 $5 $6 0 $8 $9 $10)))
                -80)
         (TRUE (VAL (PORT BUS-DATA PA)
                    (N (1 1 1 1 1 1 1 1) (1 1 1 1 1 1 1 1)))
                20)
         (TRUE (VAL (PORT BUS-ADDRESS PA)
                    (N (1 1 0 1 1 1 1 $222 1 0)
                       (1 1 0 1 1 1 1 $222 1 0)))
                0)
         (TRUE (VAL (PORT BUS-CONTROL PA)
                    (N ($223 $224 1 $225 $226 $227)
                       ($223 $224 1 $225 $226 $227)))
                0))
    (TRUE (VAL (PORT PBUS-CONTROL PA) (D (1 0 1 1) (0 1 0 0))) 160))
```

Trying to achieve test:

```
(IF (TRUE (VAL (PORT CONTROL PA-DECODER)
               (N ($223 $224 $237 1 $226 $227)
                  ($223 $224 $237 1 $226 $227)))
          0)
    (TRUE (VAL (PORT WCONTROL PA-DECODER) (D 1 0)) 100))
```

Solution:

```
(IF (AND (TRUE (VAL (PORT BUS-ADDRESS PA)
                    (N ($12 $2 $3 $4 $5 $6 0 $8 $9 $10)
                       ($12 $2 $3 $4 $5 $6 0 $8 $9 $10)))
               -80)
         (TRUE (VAL (PORT BUS-DATA PA)
                    (N (1 1 1 1 1 1 1 1) (1 1 1 1 1 1 1 1)))
               20)
         (TRUE (VAL (PORT BUS-CONTROL PA)
                    (N ($15 $16 $20 0 $18 $19)
                       ($15 $16 $20 0 $18 $19)))
               -40)
         (TRUE (VAL (PORT BUS-ADDRESS PA)
                    (N (1 1 0 1 1 1 1 $8 1 0)
                       (1 1 0 1 1 1 1 $8 1 0)))
               -40)
         (TRUE (VAL (PORT BUS-RESET PA) (N 1 1)) 60)
         (TRUE (VAL (PORT BUS-CONTROL PA)
                    (N ($223 $224 1 $225 $226 $227)
                       ($223 $224 1 $225 $226 $227)))
               0))
    (TRUE (VAL (PORT PBUS-CONTROL PA) (D (0 1 0 0) (1 0 1 1))) 160))
```

Trying to achieve test:

```
(IF (TRUE (VAL (PORT ADDRESS PA-DECODER)
               (N ($28 $29 $30 $31 $32 $33 $34 $22 $35 1)
                  ($28 $29 $30 $31 $32 $33 $34 $22 $35 1)))
          0)
    (TRUE (VAL (PORT WCONTROL PA-DECODER) (D 1 0)) 100))
```

Solution:

```
(IF (AND (TRUE (VAL (PORT BUS-ADDRESS PA)
                    (N ($12 $2 $3 $4 $5 $6 0 $8 $9 $10)
                       ($12 $2 $3 $4 $5 $6 0 $8 $9 $10)))
               -80)
         (TRUE (VAL (PORT BUS-DATA PA)
                    (N (1 1 1 1 1 1 1 1) (1 1 1 1 1 1 1 1)))
               20)
         (TRUE (VAL (PORT BUS-CONTROL PA)
                    (N ($15 $16 $20 0 $18 $19)
                       ($15 $16 $20 0 $18 $19)))
               -40)
         (TRUE (VAL (PORT BUS-ADDRESS PA)
                    (N (1 1 0 1 1 1 1 $8 1 0)
                       (1 1 0 1 1 1 1 $8 1 0)))
               -40)
         (TRUE (VAL (PORT BUS-RESET PA) (N 1 1)) 60)
         (TRUE (VAL (PORT BUS-ADDRESS PA)
                    (N ($28 $29 $30 $31 $32 $33 $34 $22 $35 1)
                       ($28 $29 $30 $31 $32 $33 $34 $22 $35 1))) 0))
    (TRUE (VAL (PORT PBUS-CONTROL PA) (D (0 1 0 0) (1 0 1 1))) 160))
```

```
Generating tests for the receiver transmitter:

Trying to achieve test:

(IF (AND (TRUE (VAL (PORT DIR PA-RECTRAN) (N 1 1)) 0)
         (TRUE (VAL (PORT DATA1 PA-RECTRAN)
                    (N (1 0 1 0 1 0 1 0) (1 0 1 0 1 0 1 0)))
               0))
    (TRUE (VAL (PORT DATA2 PA-RECTRAN)
               (D (1 0 1 0 1 0 1 0) (1 1 1 1 1 1 1 1)))
          30))

Solution:

(IF (AND (TRUE (VAL (PORT BUS-DATA PA)
                    (N (1 0 1 0 1 0 1 0) (1 0 1 0 1 0 1 0)))
               0)
         (TRUE (VAL (PORT BUS-ADDRESS PA)
                    (N ($12 $2 $3 $4 $5 $6 0 $8 $9 $10)
                       ($12 $2 $3 $4 $5 $6 0 $8 $9 $10)))
               -100)
         (TRUE (VAL (PORT BUS-ADDRESS PA)
                    (N (1 1 0 1 1 1 1 $8 $9 $10)
                       (1 1 0 1 1 1 1 $8 $9 $10)))
               -70)
         (TRUE (VAL (PORT BUS-CONTROL PA)
                    (N ($15 $16 0 $17 $18 $19)
                       ($15 $16 0 $17 $18 $19)))
               -70))
    (TRUE (VAL (PORT BUS-DATA PA)
               (D (1 0 1 0 1 0 1 0) (1 1 1 1 1 1 1 1)))
          60))

Trying to achieve test:

(IF (AND (TRUE (VAL (PORT DIR PA-RECTRAN) (N 1 1)) 0)
         (TRUE (VAL (PORT DATA1 PA-RECTRAN)
                    (N (0 1 0 1 0 1 0 1) (0 1 0 1 0 1 0 1)))
               0))
    (TRUE (VAL (PORT DATA2 PA-RECTRAN)
               (D (0 1 0 1 0 1 0 1) (1 1 1 1 1 1 1 1)))
          30))

Solution:

(IF (AND (TRUE (VAL (PORT BUS-ADDRESS PA)
                    (N ($12 $2 $3 $4 $5 $6 0 $8 $9 $10)
                       ($12 $2 $3 $4 $5 $6 0 $8 $9 $10)))
               -100)
         (TRUE (VAL (PORT BUS-DATA PA)
                    (N (0 1 0 1 0 1 0 1) (0 1 0 1 0 1 0 1)))
               0)
         (TRUE (VAL (PORT BUS-ADDRESS PA)
                    (N (1 1 0 1 1 1 1 $8 $9 $10)
                       (1 1 0 1 1 1 1 $8 $9 $10)))
               -70)
         (TRUE (VAL (PORT BUS-CONTROL PA)
                    (N ($15 $16 0 $17 $18 $19)
                       ($15 $16 0 $17 $18 $19)))
               -70))
    (TRUE (VAL (PORT BUS-DATA PA)
               (D (0 1 0 1 0 1 0 1) (1 1 1 1 1 1 1 1)))
```

```
        60))
```

```
Trying to achieve test:
```

```
(IF (AND (TRUE (VAL (PORT DIR PA-RECTRAN) (N 0 0)) 0)
         (TRUE (VAL (PORT DATA2 PA-RECTRAN)
                    (N (1 0 1 0 1 0 1 0) (1 0 1 0 1 0 1 0)))
               0))
    (TRUE (VAL (PORT DATA1 PA-RECTRAN)
               (D (1 0 1 0 1 0 1 0) (1 1 1 1 1 1 1 1)))
          30))
```

```
Solution:
```

```
(IF (AND (TRUE (VAL (PORT BUS-CONTROL PA)
                    (N ($15 $16 0 $17 $18 $19)
                       ($15 $16 0 $17 $18 $19)))
               -100)
         (TRUE (VAL (PORT BUS-ADDRESS PA)
                    (N (1 1 0 1 1 1 1 $8 $9 $10)
                       (1 1 0 1 1 1 1 $8 $9 $10)))
               -100)
         (TRUE (VAL (PORT BUS-ADDRESS PA)
                    (N ($12 $2 $3 $4 $5 0 $7 $8 $9 $10)
                       ($12 $2 $3 $4 $5 0 $7 $8 $9 $10)))
               -130)
         (TRUE (VAL (PORT BUS-DATA PA)
                    (N (1 0 1 0 1 0 1 0) (1 0 1 0 1 0 1 0)))
               -30))
    (TRUE (VAL (PORT BUS-DATA PA)
               (D (1 0 1 0 1 0 1 0) (1 1 1 1 1 1 1 1)))
          30))
```

```
Trying to achieve test:
```

```
(IF (AND (TRUE (VAL (PORT DIR PA-RECTRAN) (N 0 0)) 0)
         (TRUE (VAL (PORT DATA2 PA-RECTRAN)
                    (N (0 1 0 1 0 1 0 1) (0 1 0 1 0 1 0 1)))
               0))
    (TRUE (VAL (PORT DATA1 PA-RECTRAN)
               (D (0 1 0 1 0 1 0 1) (1 1 1 1 1 1 1 1)))
          30))
```

```
Solution:
```

```
(IF (AND (TRUE (VAL (PORT BUS-CONTROL PA)
                    (N ($15 $16 0 $17 $18 $19)
                       ($15 $16 0 $17 $18 $19)))
               -100)
         (TRUE (VAL (PORT BUS-ADDRESS PA)
                    (N (1 1 0 1 1 1 1 $8 $9 $10)
                       (1 1 0 1 1 1 1 $8 $9 $10)))
               -100)
         (TRUE (VAL (PORT BUS-ADDRESS PA)
                    (N ($12 $2 $3 $4 $5 $6 0 $8 $9 $10)
                       ($12 $2 $3 $4 $5 $6 0 $8 $9 $10)))
               -130)
         (TRUE (VAL (PORT BUS-DATA PA)
                    (N (0 1 0 1 0 1 0 1) (0 1 0 1 0 1 0 1)))
               -30))
    (TRUE (VAL (PORT BUS-DATA PA)
               (D (0 1 0 1 0 1 0 1) (1 1 1 1 1 1 1 1))) 30))
```

```
Generating tests for the data buffer:

Trying to achieve test:

(IF (AND (TRUE (VAL (PORT ENABLE PA-BUSDBUFF) (N 0 0)) 0)
         (TRUE (VAL (PORT DATA1 PA-BUSDBUFF)
                    (N (1 0 1 0 1 0 1 0) (1 0 1 0 1 0 1 0)))
               0))
    (TRUE (VAL (PORT DATA2 PA-BUSDBUFF)
               (D (1 0 1 0 1 0 1 0) (1 1 1 1 1 1 1 1)))
          30))

Solution:

(IF (AND (TRUE (VAL (PORT BUS-CONTROL PA)
                    (N ($15 $16 0 $17 $18 $19)
                       ($15 $16 0 $17 $18 $19)))
               -100)
         (TRUE (VAL (PORT BUS-ADDRESS PA)
                    (N (1 1 0 1 1 1 1 $8 0 0)
                       (1 1 0 1 1 1 1 $8 0 0)))
               -100)
         (TRUE (VAL (PORT BUS-CONTROL PA)
                    (N ($15 $16 0 $17 $18 $19)
                       ($15 $16 0 $17 $18 $19)))
               -60)
         (TRUE (VAL (PORT BUS-ADDRESS PA)
                    (N (1 1 0 1 1 1 1 $8 $9 $10)
                       (1 1 0 1 1 1 1 $8 $9 $10)))
               -60)
         (TRUE (VAL (PORT BUS-ADDRESS PA)
                    (N ($12 $2 $3 $4 $5 $6 0 $8 $9 $10)
                       ($12 $2 $3 $4 $5 $6 0 $8 $9 $10)))
               -130)
         (TRUE (VAL (PORT BUS-ADDRESS PA)
                    (N (1 1 0 1 1 1 1 $8 0 0)
                       (1 1 0 1 1 1 1 $8 0 0)))
               -131)
         (TRUE (VAL (PORT BUS-CONTROL PA)
                    (N ($15 $16 $20 0 $18 $19)
                       ($15 $16 $20 0 $18 $19)))
               -131)
         (TRUE (VAL (PORT BUS-ADDRESS PA)
                    (N ($12 $2 $3 $4 $5 $6 0 $8 $9 $10)
                       ($12 $2 $3 $4 $5 $6 0 $8 $9 $10)))
               -180)
         (TRUE (VAL (PORT BUS-DATA PA)
                    (N (1 0 1 0 1 0 1 0) (1 0 1 0 1 0 1 0)))
               -80))
    (TRUE (VAL (PORT BUS-DATA PA)
               (D (1 0 1 0 1 0 1 0) (1 1 1 1 1 1 1 1)))
          70))

Trying to achieve test:

(IF (AND (TRUE (VAL (PORT ENABLE PA-BUSDBUFF) (N 0 0)) 0)
         (TRUE (VAL (PORT DATA1 PA-BUSDBUFF)
                    (N (0 1 0 1 0 1 0 1) (0 1 0 1 0 1 0 1)))
               0))
    (TRUE (VAL (PORT DATA2 PA-BUSDBUFF)
               (D (0 1 0 1 0 1 0 1) (1 1 1 1 1 1 1 1)))
          30))
```

Solution:

```
(IF (AND (TRUE (VAL (PORT BUS-CONTROL PA)
                    (N ($15 $16 0 $17 $18 $19)
                       ($15 $16 0 $17 $18 $19)))
               -100)
         (TRUE (VAL (PORT BUS-ADDRESS PA)
                    (N (1 1 0 1 1 1 1 $8 0 0)
                       (1 1 0 1 1 1 1 $8 0 0)))
               -100)
         (TRUE (VAL (PORT BUS-CONTROL PA)
                    (N ($15 $16 0 $17 $18 $19)
                       ($15 $16 0 $17 $18 $19)))
               -60)
         (TRUE (VAL (PORT BUS-ADDRESS PA)
                    (N (1 1 0 1 1 1 1 $8 $9 $10)
                       (1 1 0 1 1 1 1 $8 $9 $10)))
               -60)
         (TRUE (VAL (PORT BUS-ADDRESS PA)
                    (N ($12 $2 $3 $4 $5 $6 0 $8 $9 $10)
                       ($12 $2 $3 $4 $5 $6 0 $8 $9 $10)))
               -130)
         (TRUE (VAL (PORT BUS-ADDRESS PA)
                    (N (1 1 0 1 1 1 1 $8 0 0)
                       (1 1 0 1 1 1 1 $8 0 0)))
               -131)
         (TRUE (VAL (PORT BUS-CONTROL PA)
                    (N ($15 $16 $20 0 $18 $19)
                       ($15 $16 $20 0 $18 $19)))
               -131)
         (TRUE (VAL (PORT BUS-DATA PA)
                    (N (0 1 0 1 0 1 0 1) (0 1 0 1 0 1 0 1)))
               -80)
         (TRUE (VAL (PORT BUS-ADDRESS PA)
                    (N ($12 $2 $3 $4 $5 $6 0 $8 $9 $10)
                       ($12 $2 $3 $4 $5 $6 0 $8 $9 $10)))
               -180))
    (TRUE (VAL (PORT BUS-DATA PA)
               (D (0 1 0 1 0 1 0 1) (1 1 1 1 1 1 1 1)))
          70))
```

```
Generating tests for the control buffer:

Trying to achieve test:

(IF (AND (TRUE (VAL (PORT IRQ PA-BUSCBUFF) (N 1 1)) 0)
         (TRUE (VAL (PORT STATUS PA-BUSCBUFF)
                    (N ($222 0 $223 $224 $225)
                       ($222 0 $223 $224 $225)))
               0))
    (TRUE (VAL (PORT IRQ7 PA-BUSCBUFF) (D 1 0)) 60))

Solution:

(IF (AND (TRUE (VAL (PORT PBUS-STATUS PA)
                    (N ($222 0 $223 $224 $225)
                       ($222 0 $223 $224 $225)))
               0)
         (TRUE (VAL (PORT BUS-ADDRESS PA)
                    (N ($12 $2 $3 $4 $5 $6 0 $8 $9 $10)
                       ($12 $2 $3 $4 $5 $6 0 $8 $9 $10)))
               -130)
         (TRUE (VAL (PORT BUS-ADDRESS PA)
                    (N (1 1 0 1 1 1 1 $8 1 0)
                       (1 1 0 1 1 1 1 $8 1 0)))
               -131)
         (TRUE (VAL (PORT BUS-CONTROL PA)
                    (N ($15 $16 $20 0 $18 $19)
                       ($15 $16 $20 0 $18 $19)))
               -131)
         (TRUE (VAL (PORT BUS-RESET PA) (N 0 0)) -30)
         (TRUE (VAL (PORT BUS-DATA PA)
                    (N ($50 $51 1 $53 $54 $55 $56 $57)
                       ($50 $51 1 $53 $54 $55 $56 $57)))
               -70)
         (TRUE (VAL (PORT BUS-ADDRESS PA)
                    (N ($12 $2 $3 $4 $5 $6 0 $8 $9 $10)
                       ($12 $2 $3 $4 $5 $6 0 $8 $9 $10)))
               -170))
    (TRUE (VAL (PORT BUS-IR PA) (D 1 0)) 60))

Trying to achieve test:

(IF (TRUE (VAL (PORT IRQ PA-BUSCBUFF) (N 0 0)) 0)
    (TRUE (VAL (PORT IRQ7 PA-BUSCBUFF) (D 0 1)) 60))

Solution:

(IF (TRUE (VAL (PORT BUS-RESET PA) (N 1 1)) -30)
    (TRUE (VAL (PORT BUS-IR PA) (D 0 1)) 60))

Trying to achieve test:

(IF (AND (TRUE (VAL (PORT ENABLE1 PA-BUSCBUFF) (N 0 0)) 0)
         (TRUE (VAL (PORT STATUS PA-BUSCBUFF)
                    (N (1 1 0 1 0) (1 1 0 1 0)))
               0))
    (TRUE (VAL (PORT DATA PA-BUSCBUFF)
               (D (0 1 0 1 0 1 1 1) (1 1 1 1 1 1 1 1)))
          60))

Solution:
```

```
(IF (AND (TRUE (VAL (PORT PBUS-STATUS PA) (N (1 1 0 1 0) (1 1 0 1 0)))
                0)
         (TRUE (VAL (PORT BUS-CONTROL PA)
                     (N ($15 $16 0 $17 $18 $19)
                        ($15 $16 0 $17 $18 $19)))
                -100)
         (TRUE (VAL (PORT BUS-ADDRESS PA)
                     (N (1 1 0 1 1 1 1 $8 0 1)
                        (1 1 0 1 1 1 1 $8 0 1)))
                -100)
         (TRUE (VAL (PORT BUS-ADDRESS PA)
                     (N (1 1 0 1 1 1 1 $8 $9 $10)
                        (1 1 0 1 1 1 1 $8 $9 $10)))
                -20)
         (TRUE (VAL (PORT BUS-CONTROL PA)
                     (N ($15 $16 0 $17 $18 $19)
                        ($15 $16 0 $17 $18 $19)))
                -20))
    (TRUE (VAL (PORT BUS-DATA PA)
                (D (0 1 0 1 0 1 1 1) (1 1 1 1 1 1 1 1)))
          110))
```

Trying to achieve test:

```
(IF (AND (TRUE (VAL (PORT ENABLE2 PA-BUSCBUFF) (N 0 0)) 0)
         (TRUE (VAL (PORT CONTROL PA-BUSCBUFF)
                     (N (0 0 0 1) (0 0 0 1)))
                0)
         (TRUE (VAL (PORT IRQ PA-BUSCBUFF) (N 0 0)) 0))
    (TRUE (VAL (PORT DATA PA-BUSCBUFF)
                (D (1 1 1 0 1 0 1 0) (1 1 1 1 1 1 1 1)))
          60))
```

Solution:

```
(IF (AND (TRUE (VAL (PORT BUS-RESET PA) (N 1 1)) -30)
         (TRUE (VAL (PORT BUS-CONTROL PA)
                     (N ($15 $16 0 $17 $18 $19)
                        ($15 $16 0 $17 $18 $19)))
                -100)
         (TRUE (VAL (PORT BUS-ADDRESS PA)
                     (N (1 1 0 1 1 1 1 $8 1 0)
                        (1 1 0 1 1 1 1 $8 1 0)))
                -100)
         (TRUE (VAL (PORT BUS-CONTROL PA)
                     (N ($15 $16 0 $17 $18 $19)
                        ($15 $16 0 $17 $18 $19)))
                -30)
         (TRUE (VAL (PORT BUS-ADDRESS PA)
                     (N (1 1 0 1 1 1 1 $8 $9 $10)
                        (1 1 0 1 1 1 1 $8 $9 $10)))
                -30)
         (TRUE (VAL (PORT BUS-ADDRESS PA)
                     (N ($12 $2 $3 $4 $5 $6 0 $8 $9 $10)
                        ($12 $2 $3 $4 $5 $6 0 $8 $9 $10)))
                -160)
         (TRUE (VAL (PORT BUS-ADDRESS PA)
                     (N (1 1 0 1 1 1 1 $8 1 0)
                        (1 1 0 1 1 1 1 $8 1 0)))
                -161)
         (TRUE (VAL (PORT BUS-CONTROL PA)
                     (N ($15 $16 $20 0 $18 $19)
                        ($15 $16 $20 0 $18 $19)))
```

```
            -161)
      (TRUE (VAL (PORT BUS-RESET PA) (N 0 0)) -60)
      (TRUE (VAL (PORT BUS-ADDRESS PA)
                 (N ($12 $2 $3 $4 $5 $6 0 $8 $9 $10)
                    ($12 $2 $3 $4 $5 $6 0 $8 $9 $10)))
            -210)
      (TRUE (VAL (PORT BUS-DATA PA)
                 (N ($50 $51 $52 $53 1 0 1 0)
                    ($50 $51 $52 $53 1 0 1 0)))
            -110))
 (TRUE (VAL (PORT BUS-DATA PA)
            (D (1 1 1 0 1 0 1 0) (1 1 1 1 1 1 1 1)))
       100))
```

```
Generating tests for the control latch:

Trying to achieve test:

(IF (TRUE (VAL (PORT CLEAR PA-CLATCH) (N 1 1)) 0)
    (TRUE (VAL (PORT DATA2 PA-CLATCH) (D (0 0 0 0) (1 1 1 1))) 30))

Solution:

(IF (TRUE (VAL (PORT BUS-RESET PA) (N 1 1)) 0)
    (TRUE (VAL (PORT PBUS-CONTROL PA) (D (1 0 1 1) (0 1 0 0))) 60))

Trying to achieve test:

(IF (TRUE (VAL (PORT CLEAR PA-CLATCH) (N 1 1)) 0)
    (TRUE (VAL (PORT IRQ PA-CLATCH) (D 0 1)) 30))

Solution:

(IF (AND (TRUE (VAL (PORT PBUS-STATUS PA) (N (0 0 0 0 0) (0 0 0 0 0)))
               30)
         (TRUE (VAL (PORT BUS-RESET PA) (N 1 1)) 0))
    (TRUE (VAL (PORT BUS-IR PA) (D 0 1)) 90))

Trying to achieve test:

(IF (AND (TRUE (VAL (PORT CLOCK PA-CLATCH) (N 0 0)) 0)
         (TRUE (VAL (PORT CLOCK PA-CLATCH) (N 1 1)) 1)
         (TRUE (VAL (PORT CLEAR PA-CLATCH) (N 0 0)) 0)
         (TRUE (VAL (PORT DATA1 PA-CLATCH)
                    (N (1 0 1 0 1 0 1 0) (1 0 1 0 1 0 1 0)))
               0))
    (TRUE (VAL (PORT DATA2 PA-CLATCH) (D (1 0 1 0) (1 1 1 1))) 30))

Solution:

(IF (AND (TRUE (VAL (PORT BUS-CONTROL PA)
                    (N ($15 $16 $20 0 $18 $19)
                       ($15 $16 $20 0 $18 $19)))
               -100)
         (TRUE (VAL (PORT BUS-ADDRESS PA)
                    (N (1 1 0 1 1 1 1 $8 1 0)
                       (1 1 0 1 1 1 1 $8 1 0)))
               -100)
         (TRUE (VAL (PORT BUS-DATA PA)
                    (N (1 0 1 0 1 0 1 0) (1 0 1 0 1 0 1 0)))
               -40)
         (TRUE (VAL (PORT BUS-ADDRESS PA)
                    (N ($12 $2 $3 $4 $5 $6 0 $8 $9 $10)
                       ($12 $2 $3 $4 $5 $6 0 $8 $9 $10)))
               -140)
         (TRUE (VAL (PORT BUS-ADDRESS PA)
                    (N ($12 $2 $3 $4 $5 $6 0 $8 $9 $10)
                       ($12 $2 $3 $4 $5 $6 0 $8 $9 $10)))
               -99)
         (TRUE (VAL (PORT BUS-RESET PA) (N 0 0)) 0))
    (TRUE (VAL (PORT PBUS-CONTROL PA) (D (0 0 0 1) (0 1 0 0))) 60))

Trying to achieve test:

(IF (AND (TRUE (VAL (PORT CLOCK PA-CLATCH) (N 0 0)) 0)
         (TRUE (VAL (PORT CLOCK PA-CLATCH) (N 1 1)) 1)
```

```
        (TRUE (VAL (PORT CLEAR PA-CLATCH) (N 0 0)) 0)
        (TRUE (VAL (PORT DATA1 PA-CLATCH)
                   (N (0 1 0 1 0 1 0 1) (0 1 0 1 0 1 0 1)))
              0))
   (TRUE (VAL (PORT DATA2 PA-CLATCH) (D (1 0 1 0) (1 1 1 1))) 30))
```

```
Solution:
```

```
(IF (AND (TRUE (VAL (PORT BUS-CONTROL PA)
                    (N ($15 $16 $20 0 $18 $19)
                       ($15 $16 $20 0 $18 $19)))
               -100)
         (TRUE (VAL (PORT BUS-ADDRESS PA)
                    (N (1 1 0 1 1 1 1 $8 1 0)
                       (1 1 0 1 1 1 1 $8 1 0)))
               -100)
         (TRUE (VAL (PORT BUS-ADDRESS PA)
                    (N ($12 $2 $3 $4 $5 $6 0 $8 $9 $10)
                       ($12 $2 $3 $4 $5 $6 0 $8 $9 $10)))
               -140)
         (TRUE (VAL (PORT BUS-DATA PA)
                    (N (0 1 0 1 0 1 0 1) (0 1 0 1 0 1 0 1)))
               -40)
         (TRUE (VAL (PORT BUS-ADDRESS PA)
                    (N ($12 $2 $3 $4 $5 $6 0 $8 $9 $10)
                       ($12 $2 $3 $4 $5 $6 0 $8 $9 $10)))
               -99)
         (TRUE (VAL (PORT BUS-RESET PA) (N 0 0)) 0))
   (TRUE (VAL (PORT PBUS-CONTROL PA) (D (0 0 0 1) (0 1 0 0))) 60))
```

```
Trying to achieve test:
```

```
(IF (AND (TRUE (VAL (PORT CLOCK PA-CLATCH) (N 0 0)) 0)
         (TRUE (VAL (PORT CLOCK PA-CLATCH) (N 1 1)) 1)
         (TRUE (VAL (PORT CLEAR PA-CLATCH) (N 0 0)) 0)
         (TRUE (VAL (PORT DATA1 PA-CLATCH)
                    (N (1 0 1 0 1 0 1 0) (1 0 1 0 1 0 1 0)))
               0))
   (TRUE (VAL (PORT IRQ PA-CLATCH) (D 1 0)) 30))
```

```
Solution:
```

```
(IF (AND (TRUE (VAL (PORT PBUS-STATUS PA) (N (0 0 0 0 0) (0 0 0 0 0)))
               30)
         (TRUE (VAL (PORT BUS-CONTROL PA)
                    (N ($15 $16 $20 0 $18 $19)
                       ($15 $16 $20 0 $18 $19)))
               -100)
         (TRUE (VAL (PORT BUS-ADDRESS PA)
                    (N (1 1 0 1 1 1 1 $8 1 0)
                       (1 1 0 1 1 1 1 $8 1 0)))
               -100)
         (TRUE (VAL (PORT BUS-DATA PA)
                    (N (1 0 1 0 1 0 1 0) (1 0 1 0 1 0 1 0)))
               -40)
         (TRUE (VAL (PORT BUS-ADDRESS PA)
                    (N ($12 $2 $3 $4 $5 $6 0 $8 $9 $10)
                       ($12 $2 $3 $4 $5 $6 0 $8 $9 $10)))
               -140)
         (TRUE (VAL (PORT BUS-ADDRESS PA)
                    (N ($12 $2 $3 $4 $5 $6 0 $8 $9 $10)
                       ($12 $2 $3 $4 $5 $6 0 $8 $9 $10)))
               -99)
```

```
          (TRUE (VAL (PORT BUS-RESET PA) (N 0 0)) 0))
     (TRUE (VAL (PORT BUS-IR PA) (D 1 0)) 90))
```

Trying to achieve test:

```
(IF (AND (TRUE (VAL (PORT CLOCK PA-CLATCH) (N 0 0)) 0)
         (TRUE (VAL (PORT CLOCK PA-CLATCH) (N 1 1)) 1)
         (TRUE (VAL (PORT CLEAR PA-CLATCH) (N 0 0)) 0)
         (TRUE (VAL (PORT DATA1 PA-CLATCH)
                    (N (0 1 0 1 0 1 0 1) (0 1 0 1 0 1 0 1)))
               0))
     (TRUE (VAL (PORT IRQ PA-CLATCH) (D 0 1)) 30))
```

Solution:

```
(IF (AND (TRUE (VAL (PORT PBUS-STATUS PA) (N (0 0 0 0 0) (0 0 0 0 0)))
               30)
         (TRUE (VAL (PORT BUS-CONTROL PA)
                    (N ($15 $16 $20 0 $18 $19)
                       ($15 $16 $20 0 $18 $19)))
               -100)
         (TRUE (VAL (PORT BUS-ADDRESS PA)
                    (N (1 1 0 1 1 1 1 $8 1 0)
                       (1 1 0 1 1 1 1 $8 1 0)))
               -100)
         (TRUE (VAL (PORT BUS-ADDRESS PA)
                    (N ($12 $2 $3 $4 $5 $6 0 $8 $9 $10)
                       ($12 $2 $3 $4 $5 $6 0 $8 $9 $10)))
               -140)
         (TRUE (VAL (PORT BUS-DATA PA)
                    (N (0 1 0 1 0 1 0 1) (0 1 0 1 0 1 0 1)))
               -40)
         (TRUE (VAL (PORT BUS-ADDRESS PA)
                    (N ($12 $2 $3 $4 $5 $6 0 $8 $9 $10)
                       ($12 $2 $3 $4 $5 $6 0 $8 $9 $10)))
               -99)
         (TRUE (VAL (PORT BUS-RESET PA) (N 0 0)) 0))
     (TRUE (VAL (PORT BUS-IR PA) (D 0 1)) 90))
```

```
Generating tests for the data latch:

Trying to achieve test:

(IF (AND (TRUE (VAL (PORT CLOCK PA-DLATCH) (N 0 0)) 0)
         (TRUE (VAL (PORT CLOCK PA-DLATCH) (N 1 1)) 1)
         (TRUE (VAL (PORT DATA1 PA-DLATCH)
                    (N (1 0 1 0 1 0 1 0) (1 0 1 0 1 0 1 0)))
               0))
    (TRUE (VAL (PORT DATA2 PA-DLATCH)
               (D (1 0 1 0 1 0 1 0) (1 1 1 1 1 1 1 1)))
          30))

Solution:

(IF (AND (TRUE (VAL (PORT BUS-CONTROL PA)
                    (N ($15 $16 $20 0 $18 $19)
                       ($15 $16 $20 0 $18 $19)))
               -100)
         (TRUE (VAL (PORT BUS-ADDRESS PA)
                    (N (1 1 0 1 1 1 1 $8 0 0)
                       (1 1 0 1 1 1 1 $8 0 0)))
               -100)
         (TRUE (VAL (PORT BUS-ADDRESS PA)
                    (N ($12 $2 $3 $4 $5 $6 0 $8 $9 $10)
                       ($12 $2 $3 $4 $5 $6 0 $8 $9 $10)))
               -99)
         (TRUE (VAL (PORT BUS-DATA PA)
                    (N (1 0 1 0 1 0 1 0) (1 0 1 0 1 0 1 0)))
               -60)
         (TRUE (VAL (PORT BUS-ADDRESS PA)
                    (N ($12 $2 $3 $4 $5 $6 0 $8 $9 $10)
                       ($12 $2 $3 $4 $5 $6 0 $8 $9 $10)))
               -160))
    (TRUE (VAL (PORT PBUS-DATA PA)
               (D (1 0 1 0 1 0 1 0) (1 1 1 1 1 1 1 1)))
          30))

Trying to achieve test:

(IF (AND (TRUE (VAL (PORT CLOCK PA-DLATCH) (N 0 0)) 0)
         (TRUE (VAL (PORT CLOCK PA-DLATCH) (N 1 1)) 1)
         (TRUE (VAL (PORT DATA1 PA-DLATCH)
                    (N (0 1 0 1 0 1 0 1) (0 1 0 1 0 1 0 1)))
               0))
    (TRUE (VAL (PORT DATA2 PA-DLATCH)
               (D (0 1 0 1 0 1 0 1) (1 1 1 1 1 1 1 1)))
          30))

Solution:

(IF (AND (TRUE (VAL (PORT BUS-ADDRESS PA)
                    (N ($12 $2 $3 $4 $5 $6 0 $8 $9 $10)
                       ($12 $2 $3 $4 $5 $6 0 $8 $9 $10)))
               -99)
         (TRUE (VAL (PORT BUS-CONTROL PA)
                    (N ($15 $16 $20 0 $18 $19)
                       ($15 $16 $20 0 $18 $19)))
               -100)
         (TRUE (VAL (PORT BUS-ADDRESS PA)
                    (N (1 1 0 1 1 1 1 $8 0 0)
                       (1 1 0 1 1 1 1 $8 0 0)))
```

```
          -100)
     (TRUE (VAL (PORT BUS-DATA PA)
                (N (0 1 0 1 0 1 0 1) (0 1 0 1 0 1 0 1)))
          -60)
     (TRUE (VAL (PORT BUS-ADDRESS PA)
                (N ($12 $2 $3 $4 $5 0 $7 $8 $9 $10)
                   ($12 $2 $3 $4 $5 0 $7 $8 $9 $10)))
          -160))
 (TRUE (VAL (PORT PBUS-DATA PA)
           (D (0 1 0 1 0 1 0 1) (1 1 1 1 1 1 1 1)))
      30))
```

```
Generating tests for the driver:

Trying to achieve test:

(IF (TRUE (VAL (PORT DATA PA-CDRIVER) (N (1 0 1 0) (1 0 1 0))) 0)
    (TRUE (VAL (PORT CONTROL PA-CDRIVER) (D (0 0 0 1) (1 1 1 1))) 30))

Solution:

(IF (AND (TRUE (VAL (PORT BUS-ADDRESS PA)
                    (N (1 1 0 1 1 1 1 $8 1 0)
                       (1 1 0 1 1 1 1 $8 1 0)))
               -130)
         (TRUE (VAL (PORT BUS-CONTROL PA)
                    (N ($15 $16 $20 0 $18 $19)
                       ($15 $16 $20 0 $18 $19)))
               -130)
         (TRUE (VAL (PORT BUS-ADDRESS PA)
                    (N ($12 $2 $3 $4 $5 $6 0 $8 $9 $10)
                       ($12 $2 $3 $4 $5 $6 0 $8 $9 $10)))
               -131)
         (TRUE (VAL (PORT BUS-RESET PA) (N 0 0)) -30)
         (TRUE (VAL (PORT BUS-DATA PA)
                    (N ($50 $51 $52 $53 1 0 1 0)
                       ($50 $51 $52 $53 1 0 1 0)))
               -70)
         (TRUE (VAL (PORT BUS-ADDRESS PA)
                    (N ($12 $2 $3 $4 $5 $6 0 $8 $9 $10)
                       ($12 $2 $3 $4 $5 $6 0 $8 $9 $10)))
               -170))
    (TRUE (VAL (PORT PBUS-CONTROL PA) (D (0 0 0 1) (1 1 1 1))) 30))

Trying to achieve test:

(IF (TRUE (VAL (PORT DATA PA-CDRIVER) (N (0 1 0 1) (0 1 0 1))) 0)
    (TRUE (VAL (PORT CONTROL PA-CDRIVER) (D (1 1 1 0) (1 1 1 1))) 30))

Solution:

(IF (AND (TRUE (VAL (PORT BUS-ADDRESS PA)
                    (N (1 1 0 1 1 1 1 $8 1 0)
                       (1 1 0 1 1 1 1 $8 1 0)))
               -130)
         (TRUE (VAL (PORT BUS-CONTROL PA)
                    (N ($15 $16 $20 0 $18 $19)
                       ($15 $16 $20 0 $18 $19)))
               -130)
         (TRUE (VAL (PORT BUS-ADDRESS PA)
                    (N ($12 $2 $3 $4 $5 $6 0 $8 $9 $10)
                       ($12 $2 $3 $4 $5 $6 0 $8 $9 $10)))
               -131)
         (TRUE (VAL (PORT BUS-RESET PA) (N 0 0)) -30)
         (TRUE (VAL (PORT BUS-ADDRESS PA)
                    (N ($12 $2 $3 $4 $5 $6 0 $8 $9 $10)
                       ($12 $2 $3 $4 $5 $6 0 $8 $9 $10)))
               -170)
         (TRUE (VAL (PORT BUS-DATA PA)
                    (N ($50 $51 $52 $53 0 1 0 1)
                       ($50 $51 $52 $53 0 1 0 1)))
               -70))
    (TRUE (VAL (PORT PBUS-CONTROL PA) (D (1 1 1 0) (1 1 1 1))) 30))
```

Bibliography

[1] Aho, A. et. al., *The Design and Analysis of Computer Algorithms,* Addison-Wesley, Menlo Park, 1976.

[2] Amarel, S. "On Representations of Problems of Reasoning about Actions," from *Readings in Artificial Intelligence,* Tioga Publishing Company, Palo Alto, 1981.

[3] Barbacci, M. et. al. "Symbolic Manipulation of Computer Descriptions: The ISPS Computer Description Language," CMU-CS-79-137, Carnegie-Mellon University, Computer Science Department, August, 1979.

[4] Barrow, H. "Proving the Correctness of Digital Hardware Designs," *VLSI Design,* July 1984, pp 64-77.

[5] Barrow, H. "VERIFY: A Program for Proving Correctness of Digital Hardware Designs," *Artificial Intelligence,* vol. 24, no. 1-3, December, 1984, pp 437-491.

[6] Batali, J., and Hartheimer, A. "The Design Procedure Language Manual," AI Memo no. 598, Massachusetts Institute of Technology, September, 1980.

[7] Bobrow, D. and Stefik, M. "The LOOPS Manual," Technical Report KB-VLSI-81-13, Knowledge Systems Area, Xerox Palo Alto Research Center, 1981.

[8] Breuer, M. and Friedman, A. *Diagnosis and Reliable Design of Digital Systems,* Computer Science Press, 1976.

[9] Brown, H. Tong, C. and Foyster, G. "Palladio: An Exploratory Environment for Circuit Design," HPP-83-31, Stanford University Heuristic Programming Project, June, 1983.

[10] Bryant, R. "A Switch-level Model and Simulator for MOS Digital Systems," TR-5065, California Institute of Technology, 1983.

[11] Cheng, J. and Huang, T. "A Subgraph Isomorphism Algorithm using Resolution," *Pattern Recognition,* vol. 13, no. 5, 1981, pp 371-379.

[12] Comerford, R. and Lyman, J. "Self-Testing Special Report," *Electronics*, March 10, 1983, pp 109-124.

[13] Davis, R. "Diagnosis via Causal Reasoning: Paths of Interaction and the Locality Principle," *Proceedings of the 1983 AAAI Conference,* August 1983, pp 88-94.

[14] Davis, T. and Clark, J. "SILT: A VLSI Design Language," TR-226, Computer Systems Laboratory, Stanford University, October, 1982.

[15] de Kleer, J. "Causal and Teleological Reasoning in Circuit Recognition," AI-TR 529, Massachusetts Institute of Technology, September, 1979.

[16] de Kleer, J. "An Assumption-based TMS," to be published.

[17] Donath, W. and Hoffman, A. "Algorithms for Partitioning of Graphs and Computer Logic Based on Eigenvectors of Connection Matrices," *IBM Technical Discl. Bulletin,* vol. 15, no. 3, August, 1972.

[18] Doyle, J. "Truth Maintenance Systems for Problem Solving," AI-TR 419, Massachusetts Institute of Technology, January, 1978.

[19] Enderton, H. *A Mathematical Introduction to Logic,* Academic Press, 1972.

[20] Finger, J. and Genesereth, M. "Residue: A Deductive Approach to Design Synthesis," HPP-85-1, Stanford University Heuristic Programming Project, January, 1985.

[21] Foyster, G. "Helios: User's Manual," HPP-84-34, Stanford University Heuristic Programming Project, July, 1984.

[22] Garey, M. and Johnson, D. *Computers and Intractability: A Guide to the Theory of NP-Completeness*, W.H. Freeman and Company, 1979, pp 161-164.

[23] Genesereth, M. "The use of Design Descriptions in Automated Diagnosis," HPP-81-20, Stanford University Heuristic Programming Project, January, 1984.

[24] Genesereth, M. et. al. "The MRS Dictionary," HPP-80-24, Stanford University Heuristic programming Project, January, 1984.

[25] Goel, P. "Test Generation Cost Analysis and Projections," *Proceedings of the 17th Design Automation Conference,* June, 1980.

[26] Goel, P. "An Implicit Enumeration Algorithm to Generate Tests for Combinational Logic Circuits," *IEEE Transactions on Computers,* vol. c-30, no. 3, pp 215-222.

[27] Hill D. "Language and Environment for Multi-level Simulation," TR-185, Stanford University Computer Systems Laboratory, March, 1980.

[28] Hill, F. and Huey, B. "SCIRTSS: A Search System for Sequential Circuit Test Sequences," *IEEE Transactions on Computers,* May 1977, pp 490-502.

[29] Hoffmann, C. *Group-Theoretic Algorithms and Graph Isomorphism,* Springer-Verlag, New York, 1979.

[30] Ibarra, H. and Sahni, S. "Polynomially Complete Fault Detection Problems," IEEE Transaction on Computers, vol. C-24, no. 3, March 1976, pp 242-250.

[31] Kelly, V. and Steinberg L. "The CRITTER System: Analyzing Digital Circuits by Propagating Behaviors and Specifications," *Proceedings of the 1982 AAAI Conference,* August, 1982.

[32] Kim, J. "Exploiting Domain Knowledge in IC Cell Layout," *IEEE Design and Test of Computers,* August 1984, pp 52-65.

[33] Korf, R. "Toward a Model of Representation Changes," *Artificial Intelligence,* vol. 14, no. 1, August, 1980, pp 41-78.

[34] Kramer, G. "Employing Massive Parallelism in Digital ATPG Algorithms," *Proceedings of the 1983 IEEE International Test Conference,* IEEE Press, pp 108-114.

[35] Lai, K. "Functional Testing of Digital Systems," CMU-CS-81-148, Carnegie-Mellon University, December, 1981.

[36] Mark, G. "Parallel Testing of Non-volatile Memories," *Proceedings of the 1983 IEEE International Test Conference,* IEEE Press, pp 738-743.

[37] McCarthy, J. "A Tough Nut for Proof Procedures," AI Project Memo 16, Stanford University, 1964.

[38] McCluskey, E. "A Survey of Design for Testability Scan Techniques," *VLSI Design,* December, 1984, pp 38-61.

[39] McCluskey, E. "Minimization of Boolean Functions," *Bell System Technical Journal,* 35, no. 6, November, 1956, pp 1417-1444.

[40] Mei, K. et. al. "Partitioning of Digital Systems," TR-165, Stanford University Computer Systems Laboratory, April, 1979.

[41] Mitchell, T., Steinberg, L. and Shulman, J. "VEXED: a Knowledge-based VLSI Design Consultant," Rutgers AI/VLSI Project Working Paper no. 17, 1984.

[42] Moszkowski, B. "Reasoning about Digital Circuits," STAN-CS-83-970, Stanford University, June, 1983.

[43] Nilsson, N. *Problem Solving Methods in Artificial Intelligence,* McGraw-Hill, New York, 1971.

[44] Nilsson, N. *Principles of Artificial Intelligence,* Tioga Publishing Company, Palo Alto, 1980.

[45] Payne, T. and vanCleemput, W. "Automated Partitioning of Hierarchically Specified Digital Systems," *Proceedings of the Nineteenth Design Automation Conference,* June, 1982, pp 183-192.

[46] Pitman, K. *The Revised Maclisp Manual,* MIT/LCS/TR-295, Massachusetts Institute of Technology, May, 1983.

[47] Robinson, G. "Hitest— Intelligent Test Generation," *Proceedings of the 1983 IEEE International Test Conference,* IEEE Press, pp 311-323.

[48] Roth, J. "Diagnosis of Automata Failures: A Calculus and a Method," *IBM Journal of Research and Development,* vol. 10, pp 278-291, 1966.

[49] Russell, S. "The Complete Guide to MRS," Stanford University Heuristic Programming Project, in preparation.

[50] Shrobe, H. "Dependency Directed Reasoning for Complex Program Understanding," AI-TR 503, Massachusetts Institute of Technology, April, 1979.

[51] Singh, N. "Corona: A Language for Describing Designs," HPP-84-37, Stanford University Heuristic Programming Project, September, 1984.

[52] Singh, N. "MARS: A Multiple Abstraction Rule-Based Simulator," HPP-83-43, Stanford University Heuristic Programming Project, December, 1983.

[53] Smith, D. and Genesereth, M. "Ordering Conjunctive Queries," HPP-82-9, Stanford University Heuristic Programming Project, October, 1984.

[54] Smith, D. and Genesereth, M. "Controlling Recursive Inference," HPP-84-6, Stanford University Heuristic Programming Project, June, 1984.

[55] Stefik, M. "Planning with Constraints," *Artificial Intelligence,* vol. 16, no. 2, 1981, pp 111-139.

[56] Stefik, M., Bobrow, D., Brown, H., Conway, L., and Tong, C. "The Partitioning of Concerns in Digital System Design," *Proceedings of the Conference on Advanced Research in VLSI,* P. Penfield (Ed.), Artech House, January, 1982.

[57] Tong, C. "Knowledge-based Circuit Design," Ph.D. dissertation, Department of Computer Science, Stanford University, 1984.

[58] Ullman, J. "An Algorithm for Subgraph Isomorphism," *Journal of the ACM,* vol. 23, 1976, pp 31-42.

[59] Williams, B. "Qualitative Analysis of MOS Circuits," AI-TR 767, Massachusetts Institute of Technology, July, 1984.

[60] Wong, A. and Akinniyi, F. "A New Product Graph Based Algorithm for Subgraph Isomorphism," *Proceedings of the 1983 IEEE Computer Vision and Pattern Recognition Conference,* IEEE Press, pp 457-467.